The Ultimate Book of Rare and Unusual Knowledge

Discover Fascinating Insights About the World Around You: Captivating yet Interesting Facts and Stories About History, Science, Technology, Culture, Animals, and More for Curious Minds!

Book 6 of Eleven Worlds to Explore
Ethereal Ray

Copyright © 2024 Ethereal Ray
All rights reserved.

No part of this publication may be reproduced, stored in a retrieval system, or transmitted in any form or by any means, electronic, mechanical, photocopying, recording, or otherwise, without the prior written permission of the author, except for brief quotations in reviews or scholarly analysis.

Images/Illustration designed by Freepik

Table of Contents

Chapter 1: History's Hidden Gems 1
Discover the **shocking moments** in history that shaped our world, from **mysterious figures** to **unsolved mysteries**.

Chapter 2: Unbelievable Science............................... 16
Explore the game-changing **discoveries** that transformed our understanding of **science** and **the universe**.

Chapter 3: Technology That Changed the World............ 17
Unveil the inventions **and** innovations **that revolutionized society and shaped the modern world.**

Chapter 4: Cultural Curiosities................................43
Delve into the **art**, **traditions**, and **architecture** that define human culture and history.

Chapter 5: Mysteries and Unsolved Puzzles...................58
From majestic landscapes to **unexplained natural phenomena**, this chapter showcases nature's awe-inspiring beauty.

Chapter 6: Unexpected Natural Phenomena..................73
From glowing ocean waves to rocks that move on their own, nature is full of mysteries. Explore the strangest, most mind-boggling natural events that challenge our understanding of the world.

Chapter 7: History's Greatest Conspiracies 87
History is full of unsolved puzzles and hidden secrets. Did the Knights Templar really hide a lost treasure?

Was there a secret chamber in the Great Pyramid? Uncover the biggest historical conspiracies that still spark debate today.

Chapter 8: Breakthroughs That Changed the World...... 101
Science and technology have transformed our lives, from the discovery of DNA to artificial intelligence and space exploration. Learn about the greatest innovations that have reshaped humanity—and what the future might hold.

Chapter 9: Unbelievable Feats of Human Ingenuity 1144
Celebrate the **incredible achievements** of humanity, from **engineering marvels** to **space exploration**.

Chapter 10: Incredible Animals............................... 1300
Discover the **extraordinary creatures** and their **survival secrets** in the animal kingdom.

The Ultimate Rare and Unusual Knowledge Quiz.......... 144

Review Request from Professor Atlas – Share Your Discovery! 1573

Hi there! I'm **Professor Atlas**, your curious and knowledgeable guide on this incredible journey of discovery. I've traveled the world, studied the wonders of nature, and explored the mysteries of history, science, technology, and animals.

Together, we'll uncover fascinating facts and stories that will expand your mind and spark your curiosity!

If you enjoy this exciting adventure into the world around us, perhaps you could ask a grown-up to leave a review on **Amazon? Reviews** help us uncover even more mysteries and create more exciting books for curious readers like you!

Review us on Amazon US!

or any Amazon site where you purchased this book!

"The present is theirs; the future, for which I really worked, is mine."

- Nikola Tesla

Get ready to dive into an incredible journey through the wonders of the world! **"The Ultimate Book of Rare and Unusual Knowledge: Discover Fascinating Insights About the World Around You: Captivating yet Interesting Facts and Stories About History, Science, Technology, Culture, Animals, and More for Curious Minds!"** is packed with fascinating information, stunning visuals, and fun activities designed to ignite your curiosity and expand your understanding of the world.

This book is more than just a collection of facts—it's your passport to an unforgettable adventure:

- **Ignite your imagination**: From the deepest oceans to the tallest mountains, uncover the mysteries of the natural world, meet extraordinary creatures, and explore the wonders of human achievement. Every chapter will spark your curiosity and take you on an unforgettable journey.

- **Fuel your mind**: Learn about the incredible inventions that have changed the world, discover historical moments that reshaped societies, and delve into fascinating animal behaviors and ecosystems. This book is a knowledge treasure chest for curious minds of all ages.

- **Expand your horizons**: From the smallest atom to the vastness of the universe, uncover the secrets of the world's greatest wonders and gain a deeper understanding of the forces that shape our reality.

- **Bond through discovery**: Share this adventure with family and friends. Dive into the world's mysteries together, create fun crafts, and embark on a journey of learning and discovery.

Embark on an exploration of mind-blowing facts and stories about the world's history, science, culture, and the amazing animals that share our planet. Each page you turn brings new excitement and inspiration.

So, grab your thinking cap, get ready to dive in, and join us on a captivating adventure filled with fascinating knowledge!

Chapter 1:
History's Hidden Gems

History isn't just about dates and events—it's a treasure trove of fascinating stories, strange coincidences, and remarkable achievements.

Some tales are so incredible that they seem more like fiction than fact.

In this chapter, we'll uncover history's quirks and surprises, from eerie connections between famous figures to ancient innovations that shaped the world.

The Lincoln and Kennedy Connection

Abraham Lincoln and John F. Kennedy, two of America's most iconic presidents, share some of the most bizarre historical coincidences:

- Lincoln was elected president in 1860, Kennedy in 1960—exactly 100 years apart.
- Both were assassinated on a Friday, seated next to their wives.
- Each was succeeded by a vice president named Johnson (Andrew Johnson for Lincoln, Lyndon B. Johnson for Kennedy).
- Lincoln had a secretary named Kennedy, while Kennedy had a secretary named Lincoln.
- Lincoln's assassin, John Wilkes Booth, was born in 1839; Kennedy's assassin, Lee Harvey Oswald, was born in 1939.

Fun Fact: Both assassins were killed before they could stand trial.

2. The Titanic and the Fictional Prediction

In 1898, 14 years before the Titanic disaster, author Morgan Robertson wrote a novella titled Futility. The book described a massive, "unsinkable" ship called the Titan that hit an iceberg and sank. The eerie similarities don't stop there: Both the fictional Titan and the Titanic were British ships and sank in the North Atlantic in April. Both carried far fewer lifeboats than necessary. Both were described as the largest ships of their time.

Fun Fact: While Robertson claimed it was purely coincidental, many view his book as a haunting prophecy.

3. Mark Twain and Halley's Comet

The great American writer Mark Twain was born in 1835, the same year Halley's Comet passed Earth. Twain predicted he would die during the comet's next appearance, saying, "It will be the greatest disappointment of my life if I don't go out with Halley's Comet."

Fun Fact: True to his prediction, Twain passed away in 1910, just one day after the comet's return. To this day, his connection to the comet remains one of history's strangest coincidences.

4. The Great Emu War

In 1932, Australian farmers faced a peculiar enemy: emus. These flightless birds invaded farmland in huge numbers, destroying crops and frustrating farmers. The military was called in, armed with machine guns, to deal with the problem. The result? The emus won. Despite repeated attempts, the soldiers couldn't keep up with the emus' speed and agility.

Fun Fact: The emus' success led to Australia adopting fences as a more effective defense!

5. The Dancing Plague of 1518

In 1518, in Strasbourg, France, dozens of people began dancing uncontrollably in the streets. This wasn't a celebration—it was an uncontrollable compulsion. Some danced for days, collapsing from exhaustion or even dying.

Fun Fact: Historians speculate it may have been caused by mass hysteria, stress, or even ergot poisoning (a hallucinogenic fungus on bread). Whatever the cause, the Dancing Plague remains one of history's strangest phenomena.

6. Hatshepsut: Egypt's Queen Who Became King

Long before Cleopatra, Hatshepsut was one of ancient Egypt's most successful rulers. After her husband's death, she declared herself Pharaoh and ruled Egypt for over two decades. To solidify her authority, she often wore a false beard and adopted male titles, defying societal norms. Under her reign, Egypt prospered, with grand temples and trade expeditions that expanded its wealth and influence.

Fun Fact: Hatshepsut's temple at Deir el-Bahri is considered one of the greatest architectural marvels of ancient Egypt.

7. Tulip Mania: The First Economic Bubble

In 17th-century Holland, tulips became a symbol of wealth and status. Prices skyrocketed, with rare tulip bulbs selling for more than the cost of a house. At its peak, a single bulb could be worth the equivalent of a luxury car today.

Fun Fact: When the market inevitably collapsed, many investors lost everything. Tulip Mania remains one of history's earliest examples of an economic bubble.

8. The Antikythera Mechanism

Discovered in a shipwreck off the coast of Greece, the Antikythera Mechanism is a 2,000-year-old device often called the world's first computer. Its intricate system of gears could predict eclipses, track the positions of planets, and even follow the Olympic Games calendar.

Fun Fact: The device was likely used by ancient Greek navigators, proving their advanced understanding of astronomy.

9. Hypatia of Alexandria: A Martyr for Knowledge

In ancient Alexandria, Hypatia was a philosopher, mathematician, and astronomer who taught and debated in public forums. Her intelligence and influence made her a symbol of wisdom—but also a target. She was brutally murdered by a mob in 415 CE, a tragic reminder of the dangers of challenging societal norms.

Fun Fact: Hypatia remains an icon of intellectual freedom and courage.

10. Catherine the Great's Unusual Ice Palace

In 1740, Russian Empress Catherine the Great ordered the construction of an elaborate ice palace as part of a royal celebration. This incredible structure was built entirely of ice, from its walls to its furniture. Even the cannons used for the event were made of ice and actually worked (loaded with gunpowder).

Fun Fact: The ice palace is considered one of the most extraordinary architectural feats of its time.

11. The Explosion That Shaped a Presidency

In 1947, President Harry S. Truman narrowly avoided an assassination attempt thanks to a bizarre twist of fate. An accidental explosion during a White House renovation exposed structural weaknesses in the building, forcing Truman to relocate. It was later discovered that the would-be assassin had planned to strike while Truman was at the White House.

Fun Fact: The explosion is considered a strange stroke of luck.

12. The "Great Stink" of London

In the summer of 1858, the Thames River in London became so polluted that the stench overwhelmed the city. Known as the Great Stink, it forced Parliament to take action, leading to the construction of a modern sewer system.

Fun Fact: The smell was so bad that members of Parliament considered relocating to avoid it!

13. Cleopatra's Pearls

To demonstrate her immense wealth and power, Cleopatra once dissolved a pearl—one of the most valuable objects in the ancient world—in vinegar and drank it as part of a bet with Mark Antony. This extravagant act was meant to prove that she could host the most expensive banquet in history.

Fun Fact: Pearls were considered the ultimate luxury item in Cleopatra's time.

14. The Banquet of Chestnuts

In 1501, Pope Alexander VI hosted a scandalous party known as the Banquet of Chestnuts. Attended by nobles and clergy, it became infamous for its debauchery and extravagance, with reports of excessive indulgence and wild competitions.

Fun Fact: The event remains one of the most controversial chapters in papal history.

15. The Oldest Known Recipe

The world's oldest written recipe is for beer, dating back over 5,000 years to ancient Mesopotamia. Written in cuneiform on a clay tablet, the recipe describes a process involving fermented barley and bread.

Fun Fact: Beer was so highly valued in ancient times that it was often used as currency.

16. The Mystery of the Green Children of Woolpit

In 12th-century England, two children with green-tinted skin appeared in the village of Woolpit. They spoke an unknown language and claimed to come from a place called "St. Martin's Land," where the sun never shone.

Fun Fact: Some believe they were malnourished, while others think the story is pure folklore.

17. The Ghost Army of World War II

During World War II, the U.S. military deployed the Ghost Army, a unit of artists, actors, and engineers tasked with deceiving enemy forces. They used inflatable tanks, fake radio transmissions, and sound effects to create the illusion of massive troop movements.

Fun Fact: The Ghost Army's tactics were so convincing that they helped turn the tide of the war in several battles.

18. The Library of Ashurbanipal

Long before the Library of Alexandria, the Assyrian king Ashurbanipal amassed a collection of over 30,000 clay tablets in Nineveh. His library contained knowledge on astronomy, medicine, and even literature, including the Epic of Gilgamesh.

Fun Fact: The Epic of Gilgamesh, one of the world's oldest stories, is a legendary tale of a Sumerian king's quest for immortality.

19. The Day the Mississippi River Ran Backward

In 1811, a massive earthquake along the New Madrid fault caused the Mississippi River to flow backward temporarily. The quake was so powerful that it created new lakes and changed the landscape of the region.

Fun Fact: This earthquake remains one of the most powerful in U.S. history.

20. The Man in the Iron Mask

A mysterious prisoner known as the Man in the Iron Mask was held in French prisons during the reign of Louis XIV. His identity has never been confirmed, leading to countless theories, including that he was the king's illegitimate brother.

Fun Fact: The story inspired the famous novel by Alexandre Dumas.

21. The Boston Molasses Disaster

In 1919, a massive tank of molasses burst in Boston, sending a wave of sticky syrup through the streets at 35 miles per hour. The disaster killed 21 people and injured over 100.

Fun Fact: The cleanup took months, and on hot days, the area reportedly still smells like molasses.

22. Emperor Nero's "Crystal Palace"

After the Great Fire of Rome in 64 AD, Emperor Nero built the Domus Aurea, or "Golden House," a sprawling palace covered in gold and jewels. It even had a rotating dining room that mimicked the movement of the stars.

Fun Fact: The palace was so extravagant that it drained Rome's treasury, earning Nero a reputation for excess.

23. The Dancing Goat that Started Coffee

According to legend, a goat herder in Ethiopia noticed his goats became energetic after eating red berries from a particular plant. Curious, he tried the berries himself, leading to the discovery of coffee.

24. The City That Moved

In 1950, the entire town of Hibbing, Minnesota, was relocated two miles to access valuable iron ore deposits beneath it. Homes, buildings, and even streets were transported in one of history's largest urban moves.

25. The Tunguska Event

In 1908, a massive explosion flattened 800 square miles of Siberian forest near the Tunguska River. The cause? Likely a meteor or comet exploding in the atmosphere, though no impact crater was found.

Fun Fact: The energy released was equivalent to 1,000 atomic bombs.

26. The Tale of the Trojan Horse

The story of the Trojan Horse is one of history's most famous legends. Greek soldiers hid inside a giant wooden horse, which the Trojans unknowingly brought into their city as a "gift." That night, the soldiers emerged, opening the gates for the Greek army to claim victory in the Trojan War.

Fun Fact: Historians debate whether the Trojan Horse was a real event or a metaphor for ancient siege tactics.

27. The World's Oldest Love Poem

Dating back to 2000 BCE, the world's oldest known love poem was discovered in ancient Mesopotamia. Written on a clay tablet in cuneiform, it's a dedication to the goddess Inanna, celebrating love and passion.

28. The Gunpowder Plot of 1605

In a failed attempt to assassinate King James I, a group of English Catholics, including the infamous Guy Fawkes, plotted to blow up the Houses of Parliament. The plot was discovered, and Guy Fawkes' capture is now commemorated every year in the UK on Bonfire Night.

Fun Fact: The phrase "Remember, remember, the fifth of November" originates from this event.

29. The Invention of Zero

The concept of zero revolutionized mathematics, and it was first developed by ancient Indian mathematicians. While other civilizations used placeholders, the Indian scholar Brahmagupta formalized zero as a number around 600 CE.

Fun Fact: The invention of zero made modern arithmetic and algebra possible.

30. The Curse of Tutankhamun

When King Tutankhamun's tomb was opened in 1922, a series of mysterious deaths among those involved led to rumors of a pharaoh's curse. While most scientists attribute the deaths to natural causes like infection or exposure to ancient toxins, the legend of the curse persists.

31. The City of Pompeii

In 79 CE, the Roman city of Pompeii was buried under volcanic ash from Mount Vesuvius. The eruption preserved the city in incredible detail, providing a snapshot of life in ancient Rome.

Fun Fact: Today, visitors can see frescoes, graffiti, and even cast molds of people frozen in their final moments.

32. The Dancing Cats of the Middle Ages

In medieval Europe, cats were often associated with witchcraft. During festivals, towns would sometimes stage "cat dances," where cats were placed in bags or on leashes and forced to "dance" to music. These practices were rooted in superstition and fear.

33. The Lighthouse of Alexandria

One of the Seven Wonders of the Ancient World, the Lighthouse of Alexandria was built in Egypt around 280 BCE. Standing over 300 feet tall, it guided sailors safely into port for centuries before being destroyed by earthquakes.

Fun Fact: It's believed to have been the tallest man-made structure in the world for nearly 1,000 years.

34. The Lost Colony of Roanoke

In 1587, English settlers established the colony of Roanoke in present-day North Carolina. When supply ships returned three years later, the colony had vanished, leaving only the word

"CROATOAN" carved into a tree. This mysterious disappearance has puzzled historians for centuries, making it one of history's most enduring mysteries.

The fate of the settlers remains unknown. Theories range from integration with local Native American tribes to a devastating attack, but no definitive evidence has ever been found.

35. The Legend of Prester John

For centuries, Europeans believed in the existence of Prester John, a mythical Christian king who supposedly ruled a powerful kingdom somewhere in Asia or Africa.

Fun Fact: Explorers sought his kingdom for generations, but it was never found.

36. The Year Without a Summer

In 1816, a massive volcanic eruption of Mount Tambora in Indonesia caused a global climate anomaly, leading to the Year Without a Summer. Crops failed, and temperatures dropped worldwide, resulting in famine and social unrest.

Fun Fact: The gloomy weather inspired Mary Shelley to write Frankenstein during this time.

37. The Legend of El Dorado

The myth of El Dorado, a city of gold hidden in South America, lured explorers for centuries. Spanish conquistadors searched for this legendary treasure, but it was never found.

Fun Fact: Some believe the legend originated from indigenous rituals involving gold dust and offerings.

38. The Origins of April Fool's Day

The exact origin of April Fool's Day is unknown, but it's often linked to France switching to the Gregorian calendar in the 16th century. Those who still celebrated the new year in April were mocked as "April fools."

39. The Magellan Penguin

The Magellan penguin is named after the explorer Ferdinand Magellan, who first encountered them during his voyage around South America in the early 16th century.

Fun Fact: These small but feisty penguins are known for their loyalty, often mating for life.

40. The Longest Siege in History

The Siege of Candia, on the island of Crete, lasted from 1648 to 1669, making it the longest siege in recorded history. The Ottoman Empire fought for 21 years to capture the city from the Venetians, eventually succeeding.

Conclusion:

History is more than a collection of dates and events—it's a treasure trove of fascinating stories, unexpected connections, and moments that shaped our world in surprising ways. From ancient mysteries to uncanny coincidences, this chapter reveals how the past continues to intrigue and inspire us. As we journey through history, we uncover not just what happened, but why it matters.

Each tale reminds us that the stories of yesterday have lessons and marvels that resonate even today. Who knows what other hidden gems history still holds, waiting to be discovered?

Chapter 2:
Unbelievable Science

Science has transformed the way we understand the world, but it's not just about formulas and experiments.

It's filled with mind-blowing discoveries, accidental breakthroughs, and phenomena that defy our expectations.

In this chapter, we'll dive into the quirky, astonishing, and sometimes unbelievable side of science that has shaped our lives and challenged our imaginations.

1. Tardigrades: The Indestructible Microscopic Creatures

Tardigrades, also known as "water bears," are nearly indestructible. These tiny creatures can survive boiling heat, freezing cold, radiation, and even the vacuum of space. They can go without water for decades by entering a state called cryptobiosis, reviving when conditions improve.

Fun Fact: Tardigrades have been found on mountaintops, in deep-sea trenches.

2. Accidental Discovery of Penicillin

In 1928, Alexander Fleming returned from vacation to find that mold had killed bacteria in one of his Petri dishes. This accidental discovery led to penicillin, the world's first antibiotic, saving millions of lives.

3. The Electric Eel's Shocking Power

Electric eels can generate up to 600 volts, enough to stun prey or deter predators. They produce electricity using specialized cells called electrocytes, which work like tiny batteries.

Fun Fact: Electric eels use their shocking ability to navigate murky waters, like a biological radar.

4. Bumblebees and Aerodynamics

For years, scientists were baffled by how bumblebees could fly. Their wings seemed too small to lift their heavy bodies. It turns out their unique wing motion creates mini whirlwinds that provide lift, defying traditional aerodynamics.

5. The Double-Slit Experiment: A Quantum Puzzle

This famous experiment shows how light and particles can behave as both waves and particles. Even stranger, particles seem to change behavior when observed, raising questions about reality itself.

Fun Fact: The double-slit experiment is often referred to as the foundation of quantum mechanics.

6. The Periodic Table and Its Gaps

Dmitri Mendeleev, the creator of the periodic table, left gaps for elements that hadn't been discovered yet. Remarkably, many of these elements, like gallium and germanium, were later found and fit perfectly into his predictions.

7. Bioluminescence: Nature's Light Show

Many marine creatures, like jellyfish and plankton, produce their own light through bioluminescence. This natural glow is used for communication, camouflage, and even luring prey.

Fun Fact: Fireflies are the most well-known land creatures with bioluminescence.

8. The Mystery of Dark Matter

Scientists estimate that 85% of the universe's mass is made of dark matter, a mysterious substance that doesn't emit or absorb light. While we can't see it, we know it exists because of its gravitational effects on galaxies.

9. Rosalind Franklin's DNA Discovery

In the 1950s, Rosalind Franklin captured X-ray diffraction images of DNA, including the famous Photo 51, which revealed its double-helix structure. Her work paved the way for Watson and Crick's model of DNA.

Fun Fact: Franklin's research also advanced our understanding of viruses, aiding in vaccine development.

10. The Sound of Black Holes

In 2022, NASA released audio of sound waves from a black hole. These eerie, low-pitched noises come from vibrations in surrounding gas, translating cosmic phenomena into sound we can "hear."

11. The Upside-Down World of Snakes

Some snakes, like the Eastern hognose snake, play dead when threatened. They flip onto their backs, stick out their tongues, and even emit a foul smell to appear lifeless.

12. Saturn's Hexagon

At Saturn's north pole, there's a mysterious hexagon-shaped storm spanning thousands of miles. Scientists believe it's caused by unique jet stream patterns, but its perfect geometric shape remains a marvel.

13. Why Ice Floats

Unlike most substances, water expands when it freezes, making ice less dense than liquid water. This anomaly allows ice to float, insulating aquatic life in winter and preventing oceans from freezing solid.

14. The "Immortal" Jellyfish

The Turritopsis dohrnii jellyfish can revert to its juvenile stage after reaching adulthood, essentially starting its life cycle over. This ability makes it biologically immortal.

15. The Science of Yawning

Yawning isn't just about being tired—it's a way to cool the brain! Scientists believe yawning increases blood flow and oxygen to regulate brain temperature.

16. Radioactive Bananas

Bananas contain potassium-40, a naturally occurring isotope that makes them slightly radioactive. Don't worry, though—you'd need to eat thousands at once to feel any effects!

17. The Leap Second

To keep our clocks in sync with Earth's rotation, scientists occasionally add a leap second to the global timekeeping system. The last one was added in 2016.

18. The Voyager Probes

Launched in 1977, the Voyager probes are still traveling through space. Each carries a Golden Record with sounds and images of Earth, intended as a message to alien civilizations.

19. The Human Body's Electricity

The human body produces enough electricity to power a small light bulb. Nerve cells send electrical signals, and the heart generates electrical pulses to pump blood.

20. The Infinite Pi

Pi (3.14159...) is an irrational number, meaning it has no end or repeating pattern. Mathematicians have calculated trillions of its digits, and they're still going!

21. Exoplanets – Worlds Beyond Our Solar System

Since the discovery of the first exoplanet in 1992, scientists have identified over 5,000 exoplanets orbiting stars beyond our solar system. These distant worlds vary widely, from gas giants to rocky planets that might even harbor life.

Fun Fact: Kepler-22b, an exoplanet located 600 light-years away, is believed to have conditions suitable for life, with an Earth-like temperature range!

22. The Power of Fungi

Fungi like mycorrhizal networks form underground "internet-like" systems, connecting plants and enabling them to share nutrients and communicate.

23. The Sahara Was Once a Jungle

About 6,000 years ago, the Sahara Desert was a lush, green savanna filled with rivers and lakes. Climate shifts turned it into the vast desert we see today.

24. Isaac Newton and the Apple

Legend has it that Isaac Newton discovered gravity when an apple fell from a tree and struck him on the head. While this story might be exaggerated, Newton's work on gravitational theory revolutionized science and laid the foundation for modern physics.

Fun Fact: Newton's law of universal gravitation showed that the same force keeps the Moon in orbit and causes objects to fall to Earth.

25. The Infinite Power of the Sun

The Sun produces more energy in one second than humanity has used in its entire history. Harnessing even a fraction of this power could solve global energy needs.

26. The Higgs Boson – Unlocking the Universe's Secrets

In 2012, scientists at CERN confirmed the existence of the Higgs boson, a fundamental particle that gives mass to other particles. Dubbed the "God Particle," its discovery completed the Standard Model of Physics.

Fun Fact: The Large Hadron Collider, where the Higgs boson was discovered, is the world's largest and most powerful particle accelerator.

27. The Science of Memory – How We Remember

The human brain stores memories through a complex process involving the hippocampus. Scientists are still unraveling how we recall vivid moments and why some memories fade.

Fun Fact: Your brain has about 2.5 petabytes of storage capacity—enough to store 3 million hours of TV shows!

28. The Physics of Rainbows – Nature's Spectacle

Rainbows occur when sunlight passes through water droplets, bending and reflecting to create a spectrum of colors. This optical phenomenon is a stunning blend of light and science.

Fun Fact: A double rainbow occurs when light reflects twice inside a droplet!

29. The Discovery of Neptune

Neptune, the eighth planet, was the first to be discovered using mathematics. Astronomers predicted its existence before it was observed in 1846.

Fun Fact: Neptune's winds are the fastest in the solar system, reaching speeds of 1,200 miles per hour!

30. The Mystery of Sleep Paralysis

Sleep paralysis is a phenomenon where a person wakes up unable to move, often accompanied by hallucinations. Scientists link it to disruptions in REM sleep.

Fun Fact: Many cultures interpret sleep paralysis as encounters with spirits or demons.

31. The Chemistry of Fireworks - Exploding Colors

Fireworks get their brilliant colors from metal salts. For example, strontium produces red, while barium creates green. It's a dazzling combination of chemistry and pyrotechnics!

Fun Fact: The earliest recorded fireworks date back to 7th-century China.

32. The Discovery of Insulin - Life-Saving for Millions

In 1921, Frederick Banting and Charles Best discovered insulin, a hormone that regulates blood sugar. This breakthrough transformed the treatment of diabetes and saved countless lives.

Fun Fact: Insulin was one of the first proteins to be mass-produced using genetic engineering.

33. The Leap Year Mystery

Leap years occur every four years to keep our calendar aligned with Earth's orbit around the Sun. Without it, our seasons would slowly drift over time.

Fun Fact: The Gregorian calendar, introduced in 1582, standardized leap years.

34. The Discovery of Gravity Waves

In 2015, scientists detected gravitational waves for the first time, confirming Einstein's theory of general relativity. These ripples in spacetime occur during cosmic events like black hole collisions.

Fun Fact: Detecting gravitational waves required the most sensitive equipment ever built!

35. The Science of Photosynthesis

Plants convert sunlight into energy through photosynthesis, producing oxygen as a byproduct. This process is the foundation of life on Earth.

Fun Fact: Some algae can perform photosynthesis more efficiently than trees!

36. The World's Deepest Cave

The Veryovkina Cave in Georgia is the deepest cave on Earth, descending over 7,200 feet. Scientists explore it to study geology and subterranean ecosystems.

Fun Fact: It takes days for explorers to reach its deepest point.

37. The Golden Ratio in Nature

The Golden Ratio (1.618...) appears in nature, art, and architecture. This mathematical proportion can be seen in sunflower spirals, nautilus shells, and even galaxies.

Fun Fact: Leonardo da Vinci used the Golden Ratio in his famous works, including the Mona Lisa.

38. The Science of Eclipses

Solar and lunar eclipses occur when the Sun, Earth, and Moon align perfectly. They've fascinated and frightened humans for centuries.

Fun Fact: A total solar eclipse lasts only a few minutes but creates a once-in-a-lifetime spectacle.

39. The Discovery of Quasars

Quasars are among the brightest objects in the universe, powered by supermassive black holes at the centers of galaxies.

Fun Fact: A single quasar can outshine entire galaxies!

40. The First Artificial Satellite

In 1957, the Soviet Union launched Sputnik 1, the first artificial satellite, marking the beginning of the Space Age.

Fun Fact: Sputnik 1 sent radio signals that were detected all over the world, inspiring space exploration.

Conclusion:

Science is a gateway to understanding the extraordinary in the everyday. From the mysteries of dark matter to the wonders of bioluminescence, this chapter highlights how the pursuit of knowledge continually reshapes our world and unlocks the universe's secrets.

These incredible discoveries and phenomena remind us that science is not just about solving problems—it's about fueling curiosity, sparking innovation, and inspiring wonder

Chapter 3:
Technology That Changed the World

Technology has shaped every aspect of our lives, from the way we communicate to how we travel and even how we think.

In this chapter, we'll explore the groundbreaking inventions and technological marvels that changed the course of history.

From the wheel to the internet, these innovations not only made the world smaller but also opened new possibilities for the future.

1. The Invention of the Telephone

In 1876, Alexander Graham Bell forever changed the way humans communicated with the invention of the telephone. Before this, the only way people could communicate over long distances was by written word, which was slow and unreliable.

Fun Fact: The first telephone call Bell made was to his assistant, Thomas Watson, saying, "Mr. Watson, come here, I want to see you."

2. The First Computers

The first computers were huge, clunky machines that filled entire rooms and were used primarily for scientific calculations. One of the earliest and most famous computers was the ENIAC (Electronic Numerical Integrator and Computer), built in 1945.

Fun Fact: The ENIAC's primary purpose was to calculate artillery firing tables for the U.S. Army during World War II.

3. The Creation of the Atomic Bomb

The atomic bomb, developed during World War II by a secret project called the Manhattan Project, was the first weapon to unleash the power of the atom. The bomb's impact was profound, leading to the end of the war in the Pacific and forever altering the nature of global conflict.

Fun Fact: The first atomic bombs, "Little Boy" and "Fat Man", were dropped on Hiroshima and Nagasaki in Japan, resulting in the devastation of both cities.

4. The Birth of Artificial Intelligence

The idea of artificial intelligence (AI) has been around for centuries, but it wasn't until the mid-20th century that computer scientists began developing machines that could simulate human intelligence.

Fun Fact: In 1997, IBM's Deep Blue defeated world chess champion Garry Kasparov, marking the first time a computer defeated a human world champion in chess.

5. The Rise of the Internet

The internet has radically transformed the way we live, work, and communicate. Its origins trace back to the ARPANET project in the 1960s, funded by the U.S. Department of Defense to connect computers for military purposes.

Fun Fact: The first-ever website, info.cern.ch, is still online today, preserved by the CERN organization.

6. The Development of Airplanes

The invention of the airplane by the Wright brothers, Orville and Wilbur Wright, in 1903, marked the beginning of modern aviation. The first flight, which lasted only 12 seconds and covered 120 feet, was a breakthrough in human travel.

Fun Fact: The first commercial airliner, the Boeing 247, could carry 10 passengers and reached speeds of up to 155 miles per hour!

7. The Creation of the Lightbulb

The first artificial electric light was created in 1802 by British scientist Sir Humphry Davy. He invented the 'arc lamp'—a glowing electric arc between two carbon electrodes. However, it was impractical for everyday use, paving the way for later inventors like Edison.

Fun Fact: Edison's early lightbulbs were made using carbonized bamboo filaments, which made them burn longer and more efficiently than earlier designs.

8. The Printing Press Revolution

Invented by Johannes Gutenberg in the 15th century, the printing press was one of the most transformative advancements in human history. Using movable type, Gutenberg's machine allowed books to be mass-produced for the first time.

Fun Fact: The Gutenberg Bible, the first major book printed with the press, was completed in 1455 and remains one of the world's most valuable artifacts.

9. The Birth of the Automobile

The automobile is one of the most transformative inventions in history, reshaping cities, economies, and personal mobility. The first gasoline-powered car was built by Karl Benz in 1885, but it was Henry Ford's Model T, introduced in 1908, that revolutionized the auto industry.

Fun Fact: The Model T was originally sold for just $850, a price that dropped as production increased.

10. The Creation of the Microwave Oven

The microwave oven was invented by Percy Spencer in 1945 when he accidentally noticed that a chocolate bar in his pocket had melted after standing near a radar system.

Fun Fact: The first microwave oven, the Radarange, weighed over 750 pounds and was about 5 feet tall!

11. The Evolution of Computers: From Punch Cards to Smartphones

Computers have come a long way since the days of punch cards. Early computers like the ENIAC used punch cards to input data and were gigantic, taking up entire rooms.

Fun Fact: The first personal computer—the Altair 8800, released in 1975—was sold as a kit for $400, and buyers had to assemble it themselves.

12. The Stethoscope – Listening to the Heart of Medicine

Invented by René Laennec in 1816, the stethoscope transformed how doctors examine patients. Before its invention, physicians relied on placing their ear directly on a patient's chest to listen to heartbeats.

Fun Fact: The original stethoscope was made from a hollow wooden tube!

13. The Creation of Vaccines

In the 18th century, Edward Jenner made a groundbreaking discovery: vaccines. After observing that people who contracted cowpox (a disease in cows) did not get smallpox, he developed the first smallpox vaccine.

Fun Fact: The smallpox vaccine is credited with wiping out smallpox, making it the first human disease to be eradicated worldwide.

14. The Invention of the Camera

The invention of the camera was a game-changer for both art and history. In the early 19th century, Joseph Nicéphore Niépce captured the world's first permanent photograph in 1826.

Fun Fact: The first photograph ever taken was an image of Niépce's own backyard, and it took eight hours to capture!

15. The Invention of the Air Conditioning

Before Willis Carrier invented the modern air conditioning system in 1902, summers were uncomfortable, and many businesses shut down during the hottest months. Carrier's invention initially aimed to cool the air in a printing plant, but it quickly proved invaluable in other areas.

Fun Fact: The first air conditioners were incredibly bulky and expensive, often only found in the homes of the wealthy or in theaters and movie palaces.

16. The Steam Engine Revolution

Invented in the 18th century by James Watt, the steam engine revolutionized industries by providing efficient mechanical power. It transformed transportation, leading to the creation of steamships and railroads, and powered the Industrial Revolution.

Fun Fact: Watt's improvements made steam engines four times more efficient, sparking a global economic boom.

17. The Development of the Jet Engine

The jet engine has had a profound impact on the world, making air travel faster and more efficient. In the 1930s, Frank Whittle in the UK and Hans von Ohain in Germany independently developed the first prototypes of the jet engine.

Fun Fact: The Concorde, a supersonic jet, could travel at twice the speed of sound, cutting the flight time between New York and London to just 3.5 hours!

18. The Rise of Social Media

Social media has become an integral part of daily life, allowing individuals to connect, share, and communicate like never before. What began as simple platforms like Facebook in 2004 and Twitter in 2006 has grown into a global phenomenon.

Fun Fact: Instagram, which started as a photo-sharing app, is now a multi-billion-dollar business, with over a billion active users worldwide.

19. The Development of GPS

The Global Positioning System (GPS), originally developed by the U.S. Department of Defense in the 1970s, has become an essential tool for navigation and location tracking.

Fun Fact: There are over 30 satellites orbiting Earth, constantly sending signals to help pinpoint your location on Earth.

20. The Creation of the Hydrogen Bomb

The hydrogen bomb, developed during the Cold War, was far more powerful than the atomic bomb. It uses nuclear fusion, rather than fission, to release energy.

Fun Fact: While it was initially developed for military purposes, hydrogen bombs fundamentally changed the way nations approached nuclear power and weaponry.

21. Revolutionary Material: The First Synthetic Plastic

In 1907, Leo Baekeland created Bakelite, the first fully synthetic plastic. Resistant to heat, solvents, and electricity, it was used in everything from electrical insulators to jewelry. This marked the beginning of the modern plastics industry.

Fun Fact: Bakelite's versatility and durability made it an instant success, and it was even used in early telephones and radios.

22. Beyond Bakelite: The Rise of New Plastics

Following the invention of Bakelite, a wave of new synthetic plastics emerged. Materials like nylon, polyethylene (PE), and polyvinyl chloride (PVC) found uses in packaging, construction, and countless consumer goods, transforming industries and daily life.

Fun Fact: Nylon, invented in 1935, was first used to make toothbrush bristles and then famously in women's stockings.

23. The Invention of the Computer Mouse

In the early 1960s, Douglas Engelbart invented the computer mouse as a device to interact with a computer's graphical user interface. The first version was made of wood, with a single button.

24. The Birth of the Smartphone

The first smartphone was introduced in 1992 by IBM under the name Simon Personal Communicator. It could send emails, make phone calls, and send faxes. However, the smartphone we recognize today was popularized by Apple's iPhone in 2007.

Fun Fact: The iPhone combined phone calls, texting, internet browsing, and apps, making it the most versatile personal device.

25. The Discovery of Radio Waves

In the late 19th century, Heinrich Hertz demonstrated the existence of radio waves, electromagnetic waves capable of traveling through space and carrying sound or information. His discovery laid the foundation for the development of wireless communication.

Fun Fact: It was his successors who turned radio waves into the technology we know today.

26. The Rise of Drones

Originally used for military surveillance, drones have become increasingly popular for civilian use. With their ability to capture aerial images, monitor wildlife, deliver packages, and provide assistance in emergency situations, drones have revolutionized industries like agriculture, photography, and logistics.

Fun Fact: Drones have also been used to explore space. NASA uses drones to help with mapping planetary surfaces, and they've even been used on Mars.

27. The Development of the Laser

The invention of the laser (Light Amplification by Stimulated Emission of Radiation) in 1960 by Theodore Maiman revolutionized industries ranging from medicine to manufacturing.

Fun Fact: Lasers are also used to cut through materials like metal and glass, making them indispensable in fields like construction and medicine.

28. The Invention of the Internet Browser

Before the World Wide Web could become the giant network it is today, Tim Berners-Lee developed the first web browser. This browser allowed users to access and navigate the vast array of information on the internet.

Fun Fact: The first web browser allowed users to read text-based websites, but modern browsers have transformed how we interact with videos, images, and interactive content.

29. The Evolution of Solar Energy

Solar energy has evolved from an experimental concept to one of the world's most promising renewable energy sources. Early solar panels were inefficient and expensive, but thanks to advances in photovoltaic technology, solar power has become an affordable and efficient energy source.

Fun Fact: Solar power is used on a massive scale in solar farms, and individual solar panels are now commonly found on the roofs of homes and schools.

30. The Rise of Cryptocurrency

In 2009, Bitcoin, a decentralized digital currency, was created by the mysterious figure Satoshi Nakamoto. Bitcoin, powered by blockchain technology, has since inspired thousands of other cryptocurrencies.

Fun Fact: Bitcoin's value soared from being worth less than a dollar in its early days to over $60,000 per Bitcoin in recent years.

31. The Creation of the Electric Car

The electric car has been around since the late 19th century, but only in recent decades have electric vehicles (EVs) become mainstream. Tesla played a pivotal role in making EVs practical, while also pushing for advancements in battery technology and charging infrastructure.

Fun Fact: The Tesla Roadster, released in 2008, was the world's first highway-legal electric vehicle to use lithium-ion battery cells.

32. The Evolution of Video Games

What began in the 1950s as a simple game of Tennis for Two has become one of the largest entertainment industries in the world. The creation of arcade games like Pong in the 1970s and home consoles like the Atari 2600 in 1977 marked the beginning of the video game revolution.

Fun Fact: Super Mario Bros., released in 1985, became one of the best-selling video games of all time and is credited with saving the video game industry after the crash of 1983.

33. MRI Machines - Revolutionizing Medical Imaging

Invented in the 1970s, Magnetic Resonance Imaging (MRI) machines use powerful magnets and radio waves to create detailed images of the human body. Unlike X-rays, MRIs are non-invasive and free from radiation.

Fun Fact: The first MRI scan took almost five hours to produce one image, a process that today takes just seconds!.

34. The Invention of the Refrigerator

The refrigerator has transformed food preservation. Before its invention, people had to rely on salt, ice, or canning to preserve food. The development of the refrigeration system by Carl von Linde in the late 19th century allowed homes and businesses to keep food fresh for longer periods.

Fun Fact: The first refrigerators were large, noisy, and expensive. It wasn't until the 20th century that refrigerators became affordable and common in homes.

35. The Wright Brothers – The First Powered Flight

In 1903, Orville and Wilbur Wright made history with the first successful powered flight in Kitty Hawk, North Carolina. Their plane, the Wright Flyer, flew for 12 seconds, covering 120 feet.

Fun Fact: The Wright Flyer was made of spruce wood and muslin fabric, showcasing the simplicity of early aviation technology.

36. The Development of Wireless Communication

The invention of wireless communication changed the way we stay connected. Pioneered by Guglielmo Marconi, who developed the first practical radio, wireless technology eventually paved the way for television and cell phones.

Fun Fact: The first transatlantic radio signal was transmitted in 1901, and by the 1920s, the first commercial radio stations were broadcasting.

37. The Evolution of the Digital Camera

The digital camera has made photography more accessible and convenient. Kodak's invention of the digital sensor in the late 1980s led to the creation of the first digital cameras.

Fun Fact: The first digital camera took a blurry photo with a resolution of just 0.01 megapixels. Today, smartphone cameras can take high-resolution photos of up to 100 megapixels.

38. The Steam Turbine - Powering the Modern Age

Invented by Charles Parsons in 1884, the steam turbine revolutionized how we generate electricity. Unlike traditional steam engines, turbines convert steam into rotational energy more efficiently, becoming the backbone of power plants worldwide.

Fun Fact: Steam turbines produce over 80% of the world's electricity today, powering everything from homes to factories.

39. Global Network: The Rise of the Internet

The internet, initially a U.S. military project (ARPANET), evolved into a global network connecting billions. Tim Berners-Lee's invention of the World Wide Web in 1989 made it accessible, revolutionizing communication and information access

Fun Fact: The "@" symbol, now a universal அடையாளம் (அடையாலம் - symbol or mark) of the digital age, was chosen by email pioneer Ray Tomlinson in 1971 for its unique and underused status on the keyboard.

40. Instant Communication: The Invention of the Telegraph

Samuel Morse's invention of the telegraph in the 1830s revolutionized long-distance communication. Using electrical signals to transmit messages in Morse code, it allowed for near-instantaneous communication across continents, impacting everything from news reporting to warfare.

The first public telegraph line connected Washington, D.C., and Baltimore, Maryland, in 1844.

Conclusion:

Technology continues to shape our world in unimaginable ways, from the inventions of ancient tools to the cutting-edge gadgets of the 21st century. Each of these innovations represents a milestone in human ingenuity, bringing us closer together and allowing us to solve problems in new and exciting ways. As technology continues to evolve, who knows what groundbreaking invention will emerge next?

Chapter 4: Cultural Curiosities

Culture is a mirror reflecting the richness of human history, imagination, and creativity.

It shapes how societies function, how people express themselves, and how we understand the world.

In this chapter, we'll explore some of the most unusual, fascinating, and diverse cultural facts and practices that continue to inspire, mystify, and sometimes even baffle people across the globe.

1. The Untranslatable Words

Every language has words that can't be easily translated into another language. For example, the Japanese word "tsundoku" refers to the act of acquiring books and letting them pile up, but not reading them right away. In German, "wanderlust" refers to the deep desire to travel and explore the world.

Fun Fact: The Dutch word "gezellig" refers to a sense of coziness and contentment, often shared with others, which has no direct translation in English.

2. The Secret Language of Whistling

In the Canary Islands, there's a unique form of communication called Silbo Gomero. This language, which is based on whistling, has been used for centuries to communicate across the islands' deep ravines.

Fun Fact: Silbo Gomero has been recognized by UNESCO as an intangible cultural heritage.

3. The Mystery of the Easter Island Moai

The Moai statues of Easter Island, or Rapa Nui, are some of the most iconic and mysterious creations in the world. These massive stone heads were carved between 1400 and 1650 CE, but how the islanders transported them from quarries to their final locations remains a mystery.

Fun Fact: The Moai are thought to represent ancestral figures who had the power to ensure the well-being of the island's people.

4. The Japanese Art of Kintsugi

Kintsugi is the Japanese art of repairing broken pottery using gold, silver, or platinum. The idea is to embrace the imperfections and highlight the history of the object rather than hide the damage.

Fun Fact: Kintsugi translates to "golden joinery," and the repaired item is often considered more beautiful than before.

5. The Day of the Dead (Día de los Muertos)

Celebrated on November 1st and 2nd in Mexico, the Day of the Dead honors loved ones who have passed away. Families create elaborate altars adorned with marigolds, candles, and the favorite foods of the deceased.

Fun Fact: The marigold flower, often called the "flower of the dead," is believed to guide spirits back to the living world with its bright color and scent.

6. The Origins of Tattoos

Tattoos have been part of human culture for thousands of years. The oldest known tattoos were found on the mummified remains of Ötzi the Iceman, who lived around 5,300 years ago.

Fun Fact: The modern tattoo industry began in the early 20th century with the invention of the electromagnetic tattoo machine by Samuel O'Reilly.

7. The Mystery of the Sphinx

The Great Sphinx of Giza, a colossal limestone statue with the body of a lion and the head of a pharaoh, has puzzled historians for centuries. While its exact purpose remains unknown, it's believed to represent the pharaoh Khafre and guards the Great Pyramid.

Fun Fact: The Sphinx has no nose, and historians are still uncertain how it was destroyed.

8. The Global Influence of Ancient Egyptian Culture

Ancient Egypt has had a profound impact on world history, especially in areas like architecture, mathematics, and medicine. Their architectural feats, such as the construction of the Pyramids, are still considered some of the most impressive in human history.

Fun Fact: The ancient Egyptians were among the first to develop a 365-day calendar based on the solar year.

9. The Tradition of High Tea in England

High tea, traditionally served in England, was originally a meal for the working class, consisting of tea, sandwiches, and hearty foods like meat pies and cakes. Today, it's associated with elegance and is a beloved social tradition.

Fun Fact: The term "high tea" actually refers to a dinner meal, not a fancy afternoon snack as many people believe!

10. The Origins of the Olympic Games

The first Olympic Games were held in Ancient Greece in 776 BCE to honor the god Zeus. Athletes from all over the Greek world competed in events like foot races, wrestling, and chariot racing.

Fun Fact: The ancient Greeks only allowed men to compete, and nude athletes were a common sight during the games.

11. The Mystery of the Nazca Lines

The Nazca Lines are massive, ancient geoglyphs etched into the desert floor in Peru. These giant drawings of animals, plants, and geometric shapes can only be fully appreciated from the air.

Fun Fact: The Nazca people created the lines by removing red rocks to reveal the lighter soil underneath.

12. The Fascinating Culture of Voodoo

Voodoo is a syncretic religion with roots in West Africa, brought to the Americas by enslaved Africans. It combines elements of Catholicism and African spiritual beliefs, with practices like spirit possession and healing rituals.

Fun Fact: While Hollywood often portrays Voodoo as dark and sinister, it is actually a peaceful religion focused on healing and personal empowerment.

13. The Story of King Arthur and the Knights of the Round Table

The legend of King Arthur has been passed down through centuries of storytelling. While historians debate whether Arthur was a real person or a myth, the tales of his knights and the search for the Holy Grail have inspired countless adaptations in literature and film.

Fun Fact: The famous Excalibur sword was said to be either given to Arthur by the Lady of the Lake or pulled from a stone.

14. The Fascination with Bigfoot

Bigfoot, also known as Sasquatch, is a creature from North American folklore. Descriptions of Bigfoot vary, but the creature is usually depicted as a large, hairy, humanoid figure.

Fun Fact: The Patterson-Gimlin film, shot in 1967, is one of the most famous pieces of evidence showing Bigfoot in the wild.

15. The Symbolism of Dreamcatchers

The dreamcatcher is a Native American object traditionally used to protect individuals from bad dreams and nightmares. It consists of a circular frame with a woven net or web and feathers hanging from it.

Fun Fact: Dreamcatchers are often hung above beds, and many believe that the feathers represent the gentle breeze that carries good dreams.

16. The Origins of the Chinese Zodiac

The Chinese Zodiac is a system of assigning an animal to each year in a 12-year cycle. The cycle includes animals such as the rat, ox, tiger, rabbit, and others, each of which is associated with specific personality traits and destiny.

Fun Fact: The Chinese Zodiac also plays an important role in astrology and feng shui.

17. The Invention of the Frisbee

The Frisbee has become a symbol of outdoor fun and recreation, but it began as a pie tin. In the 1940s, college students in Connecticut started throwing pie tins from the Frisbie Pie Company and later coined the term Frisbee.

Fun Fact: The first official Frisbee disc was made by Walter Morrison and became a hit when it was produced by the Wham-O company in 1957.

18. The Tradition of Carnival in Brazil

Carnival is an annual festival held in Brazil, just before Lent. Known for its colorful costumes, lively samba music, and grand parades, Carnival is one of the biggest celebrations in the world.

Fun Fact: The Sambadrome in Rio de Janeiro is a 90,000-seat stadium built specifically for Carnival parades.

19. Samurai's Code: Honor and Discipline in Feudal Japan

The samurai, the warrior class of feudal Japan, were renowned for their martial skills and unwavering loyalty. They lived by a strict code of conduct known as Bushido, which emphasized honor, courage, and self-discipline. This code shaped not only their martial conduct but also their daily lives.

The word "samurai" itself means "to serve,

20. The Ancient Art of Falconry

Falconry, the art of training birds of prey, dates back over 3,000 years and has been practiced in cultures from the Middle East to medieval Europe. It was originally a way to hunt game.

Fun Fact: In 2010, falconry was recognized as an intangible cultural heritage by UNESCO.

21. The Art of Origami

Origami, the Japanese art of paper folding, has been practiced for centuries. It began as a ceremonial art form and has evolved into a popular hobby and art form globally.

Fun Fact: The largest origami crane ever folded was over 80 feet long and displayed in Japan as part of a peace exhibit.

22. The Mystery of the Loch Ness Monster

The Loch Ness Monster, affectionately known as Nessie, is said to inhabit Loch Ness, a large lake in Scotland. Despite numerous claims of sightings, photographs, and even sonar evidence, the monster remains elusive.

Fun Fact: The most famous image of Nessie, the "Surgeon's Photo", taken in 1934, was later revealed to be a hoax.

23. The Making of the Great Wall of China

The Great Wall of China is one of the most iconic structures in the world, stretching over 13,000 miles. Built over several dynasties, it was originally constructed to protect China from invasions.

Fun Fact: Contrary to popular belief, the wall is not visible from space with the naked eye.

24. The Secret of the Pyramids

The Egyptian pyramids remain one of the greatest architectural feats in history. Built as tombs for pharaohs, the Great Pyramid of Giza is the only remaining wonder of the ancient world.

Fun Fact: The Great Pyramid of Giza was the tallest man-made structure in the world for over 3,800 years.

25. The Surprising History of Paper

Paper was invented in China during the Han Dynasty (around 100 BCE), though its origins can be traced back even earlier with different forms of writing material. The invention of paper revolutionized communication, printing, and record-keeping.

Fun Fact: The first paper currency, also created in China, was used as a way to eliminate the need for metal coins during trade.

26. The Origin of Mardi Gras

Mardi Gras, French for "Fat Tuesday," is a colorful, exuberant festival celebrated mainly in New Orleans, Louisiana. It includes grand parades, masquerade balls, and elaborate costumes, marking the last day of revelry before the solemn season of Lent begins in Christianity.

Fun Fact: The tradition of throwing beads during Mardi Gras dates back to the 19th century when Rex, the King of Carnival, began tossing inexpensive trinkets to the crowd.

27. The Timeless Legacy of the Roman Colosseum

The Colosseum in Rome is an enduring symbol of the grandeur of the Roman Empire. Built in the 1st century AD, it was used for gladiator contests, public spectacles, and even mock sea battles.

Fun Fact: The Colosseum's underground area, known as the hypogeum, contained elevator systems and trapdoors used for dramatic entrances during gladiatorial games.

28. The Surreal Art of Salvador Dalí

Salvador Dalí, a prominent figure in the Surrealist movement, revolutionized the art world with his strange, dreamlike paintings. His most famous work, "The Persistence of Memory", features melting clocks draped across landscapes.

Fun Fact: Dalí's signature mustache was inspired by Spanish master painter Diego Velázquez and became one of his most recognizable features.

29. The Creation of the Eiffel Tower

Gustave Eiffel, a French engineer, designed the Eiffel Tower for the 1889 World's Fair in Paris. Originally criticized by many, the tower became a global icon of modernity.

Fun Fact: The Eiffel Tower was initially intended to be temporary but was saved because it proved valuable as a radio transmission tower.

30. The Oracle of Delphi – The Voice of the Gods

The Oracle of Delphi was an ancient priestess in Greece who was believed to deliver prophecies from the god Apollo. Pilgrims from across the ancient world traveled to Delphi to seek her guidance.

Fun Fact: Modern researchers believe the oracle's visions may have been caused by inhaling gases like ethylene, which leaked through cracks in the temple's floor.

31. The Symbolism of the Peace Sign

The peace sign was created in 1958 by Gerald Holtom as a symbol for the Campaign for Nuclear Disarmament (CND) in the UK. The design combines the semaphore signals for the letters N (nuclear) and D (disarmament).

Fun Fact: The peace sign was originally intended to be used exclusively for nuclear disarmament but has since become a broader symbol for peace.

32. The Art of Flamenco Dancing

Flamenco, a passionate and energetic dance form, originated in Spain, particularly in the Andalusian region. It combines singing, guitar playing, handclaps, and dancing, with deep emotional expressions.

Fun Fact: Flamenco has evolved over centuries and was influenced by Gypsy, Jewish, and Moorish cultures, making it a melting pot of musical and dance styles.

33. The History of Chocolate

Chocolate has a long history, from its origins as a bitter drink used by the ancient Mesoamerican cultures like the Aztecs and Mayans to its modern-day form as a sweet treat.

Fun Fact: The word "chocolate" comes from the Nahuatl word "xocolatl", which referred to a bitter beverage made from cacao beans.

34. The Evolution of the English Language

The English language has evolved over centuries, from Old English, which was a Germanic language, to the modern English we use today. The spread of the British Empire and the influence of Latin, French, and German have all shaped the vocabulary, spelling, and grammar.

Fun Fact: Shakespeare is credited with coining over 1,700 words in the English language.

35. The Rise of Hip-Hop Culture

Hip-hop emerged in the 1970s in New York City, blending elements of rapping, DJing, breakdancing, and graffiti. It has since become a global cultural movement, influencing everything from music and fashion to language and activism.

Fun Fact: The first hip-hop album, "Rapper's Delight" by The Sugarhill Gang, was released in 1980 and became the genre's first major commercial success.

36. The Origin of the Samurai Code of Bushido

The Bushido code is the ethical code of the samurai warriors of Japan. It emphasized loyalty, honor, courage, and self-discipline. These values not only shaped the lives of samurai but also had a profound influence on Japanese society as a whole.

Fun Fact: Bushido can be translated as "the way of the warrior," and it inspired many aspects of Japanese culture, including modern martial arts.

37. The Fascination with Mummies

The practice of mummifying bodies in ancient Egypt was part of a belief in the afterlife. The Egyptians preserved bodies to protect them from decay, allowing the spirit to live on forever.

Fun Fact: The most famous mummy, that of Tutankhamun, was discovered in 1922 by archaeologist Howard Carter and remains one of the most well-preserved mummies in history.

38. The Uniqueness of Icelandic Naming Traditions

In Iceland, surnames are typically derived from the father's first name, rather than a family name passed down through generations. For example, if a father's name is Jón, his son might have the surname Jónsson (son of Jón), and his daughter might have the surname Jónsdóttir (daughter of Jón).

Fun Fact: Icelandic naming practices are so unique that even official records are kept in first-name order rather than last-name order.

39. The Sacred Practice of Meditation

Meditation, which has been practiced for thousands of years in Buddhism, Hinduism, and other spiritual traditions, involves focusing the mind to achieve a state of mental clarity and inner peace.

Fun Fact: Meditation has been shown to have various health benefits, including reducing stress and improving concentration.

40. The Importance of the Humble Piñata

The piñata is a traditional Mexican party game that involves hitting a colorful container filled with sweets and toys. While it's now popular at birthday parties and festive celebrations around the world, the piñata has roots in ancient Aztec rituals.

Fun Fact: Traditionally, piñatas were made of clay pots, and breaking them symbolized the triumph of good over evil.

Conclusion:

From the untranslatable words that capture unique feelings to the enduring mysteries of ancient civilizations, this chapter has explored the fascinating tapestry of global cultures.

These diverse customs, traditions, and stories highlight the richness and complexity of human history. By delving into these cultural curiosities, we gain a deeper appreciation for the myriad ways in which societies around the world have shaped our present and continue to influence our lives today.

Chapter 5:
Mysteries and Unsolved Puzzles

The world is full of **unanswered questions and mysteries** that continue to intrigue and inspire us.

Whether they involve the supernatural, historical events, or scientific phenomena, some puzzles remain unsolved despite the efforts of experts.

In this chapter, we'll explore some of the most famous and curious mysteries that have puzzled humanity for centuries.

1. The Bermuda Triangle

The Bermuda Triangle, often called the Devil's Triangle, is an area in the Atlantic Ocean bounded by Miami, Bermuda, and Puerto Rico. Over the decades, there have been numerous reports of ships and airplanes disappearing in this region under mysterious circumstances.

Fun Fact: Despite the myths, the Bermuda Triangle has been found to have no higher incidence of accidents than other heavily traveled parts of the world.

2. The Disappearance of the Roanoke Colony

The Roanoke Colony, established in 1587 on Roanoke Island (modern-day North Carolina), was the first English settlement in America. When a supply ship returned three years later, the colony had vanished without a trace, and the only clue was the word "CROATOAN" carved into a tree.

Fun Fact: The name "CROATOAN" referred to an island off the coast, home to the Croatoan tribe, leading some historians to believe the colonists may have relocated there.

3. The Lost City of Atlantis

The Lost City of Atlantis was first mentioned by Plato in his writings, where he described it as a mighty and technologically advanced civilization that existed about 9,000 years ago before sinking into the ocean.

Fun Fact: Atlantis has captured the imagination of countless writers and thinkers, influencing works in fiction, including Jules Verne's 20,000 Leagues Under the Sea.

4. The Piri Reis Map: A Cartographic Enigma

Created by Ottoman admiral Piri Reis in 1513, this map is part of a larger world map and has baffled experts for centuries. It shows the South American coastline with surprising accuracy and even hints at the existence of Antarctica, a continent not officially discovered until the 1820s.

Fun Fact: Piri Reis claimed to have drawn his map from older sources, some dating back to the time of Alexander the Great, adding to the intrigue.

5. Stonehenge

Stonehenge, located in Wiltshire, England, is an ancient stone circle that has fascinated archaeologists, astronomers, and historians for centuries. It is believed to have been constructed between 3000 BCE and 2000 BCE. The stones, some of which weigh over 25 tons, were carefully arranged to align with celestial events such as the summer solstice.

Fun Fact: It's speculated that the stones may have been transported from a quarry over 150 miles away, and there's still no definitive explanation for how they were lifted and positioned.

6. The Voynich Manuscript: Still Uncracked

The Voynich Manuscript, a 15th-century book filled with an unknown script and strange illustrations, remains undeciphered. Despite countless attempts by cryptographers and linguists, its meaning and purpose are still a mystery.

Fun Fact: The manuscript's illustrations include plants that don't match any known species, adding to its enigmatic nature.

7. Vanished Hijacker: The D.B. Cooper Mystery

In 1971, a man using the alias D.B. Cooper hijacked a plane, extorted $200,000, and then parachuted into the night, never to be seen again. Despite an extensive FBI investigation, his true identity and fate remain unknown.

Fun Fact: A small portion of the ransom money was found in 1980, but the rest, and Cooper himself, have never been found.

8. Lost City Found: The Fate of Helike

Helike, once a prosperous Greek city, was swallowed by the sea in 373 BCE after a devastating earthquake and tsunami. Long considered a myth, underwater archaeological discoveries in recent years have confirmed its existence and revealed a city frozen in time.

Fun Fact: The fate of Helike may have contributed to the legend of Atlantis, another city lost to the waves.

9. The Zodiac Killer

The Zodiac Killer terrorized northern California in the 1960s and 1970s, committing a series of murders and sending cryptic letters to the media. The killer's identity was never uncovered, and his last confirmed murder occurred in 1969.

Fun Fact: In 2021, a DNA breakthrough led to new hope for solving the case, but the killer's identity is still unknown.

10. The Mystery of the Antikythera Mechanism

Discovered in 1901 in a shipwreck off the coast of Greece, the Antikythera Mechanism is an ancient analog computer used to predict astronomical positions and eclipses. The mechanism's intricate gears and dials suggest it was highly advanced for its time, dating back to 150 BCE.

Fun Fact: The Antikythera Mechanism is one of the earliest known analog computers and is often referred to as the world's first computer.

11. The Mystery of the Black Dahlia

The Black Dahlia refers to the 1947 murder of Elizabeth Short, an aspiring actress whose body was found mutilated in a gruesome and infamous crime in Los Angeles. Despite multiple investigations and hundreds of suspects, the case remains unsolved.

Fun Fact: The Black Dahlia murder case has inspired countless books, movies, and theories, but no definitive solution has ever been found.

12. The Disappearance of Amelia Earhart

Amelia Earhart, one of the first women to fly solo across the Atlantic Ocean, disappeared in 1937 while attempting to circumnavigate the globe. Earhart's plane vanished over the Pacific Ocean, and despite extensive searches, neither her body nor her aircraft were found.

Fun Fact: Earhart's disappearance remains one of the greatest mysteries in aviation history.

13. The Mystery of the Lost City of Z

The Lost City of Z is a legendary city hidden deep in the Amazon rainforest, first proposed by British explorer Percy Harrison Fawcett in the early 20th century. Fawcett disappeared while searching for the city in 1925, and his fate remains unknown.

Fun Fact: Fawcett's disappearance sparked numerous expeditions into the Amazon, but the mystery of the Lost City of Z persists.

14. The Sodder Children: Vanished in Flames?

On Christmas Eve 1945, the Sodder family home in West Virginia burned to the ground. Five of the ten children were unaccounted for, but no remains were ever found in the ashes. Did they perish in the fire, or did something else happen that night?
Fun Fact: The Sodder family maintained a billboard for decades, offering a reward for information about their missing children.

15. The Tunguska Blast: Siberia's Unsolved Explosion

In 1908, a massive explosion rocked a remote region of Siberia near the Tunguska River. It flattened trees for miles, yet left no crater. The cause remains a mystery, though the leading theory is the airburst of a comet or meteoroid.

Fun Fact: The Tunguska event is the largest impact event on

Earth in recorded history.

16. The Search for Extraterrestrial Life

Despite decades of searching, scientists have yet to find definitive proof of extraterrestrial life. While there are many theories about life beyond Earth, including the discovery of exoplanets in the habitable zone, the question remains unanswered.

Fun Fact: The SETI (Search for Extraterrestrial Intelligence) program continues to scan the skies, hoping to detect signals from intelligent civilizations.

17. The Lost Tomb of Alexander the Great

Alexander the Great, one of the most famous military leaders in history, died in 323 BCE, but the location of his tomb remains a mystery. His tomb was said to be located in Alexandria, Egypt, but despite multiple searches over the centuries, no one has been able to find it.

Fun Fact: Some believe his tomb may still be hidden beneath the sands of Egypt.

18. The Disappearance of Flight MH370

In 2014, Malaysia Airlines Flight MH370 mysteriously disappeared while en route from Kuala Lumpur to Beijing. Despite extensive search efforts across vast areas of the Indian Ocean, the wreckage and the cause of the disappearance remain unknown.

Fun Fact: The disappearance of MH370 remains one of the most mysterious aviation incidents in modern history, with only a few pieces of wreckage confirmed to be from the flight.

19. The Mystery of the Shroud of Turin

The Shroud of Turin is a linen cloth believed by some to be the burial shroud of Jesus Christ, bearing a faint image of a man. The origins of the cloth and how the image was created remain a subject of debate.

Fun Fact: Despite multiple scientific studies, including carbon dating and X-ray fluorescence, the Shroud of Turin continues to provoke controversy and curiosity.

20. The Mysterious Death of Edgar Allan Poe

The famous writer Edgar Allan Poe died under strange circumstances in 1849. Found delirious and wearing someone else's clothes, Poe was taken to a hospital, where he died a few days later.

Fun Fact: Poe's mysterious final words were reportedly, "Lord help my poor soul."

21. The Haunted Winchester Mystery House

The Winchester Mystery House in San Jose, California, is a sprawling mansion known for its oddities. Built by Sarah Winchester, the widow of the inventor of the rifle, the house contains staircases leading to nowhere, secret passageways, and doors that open into walls.

Fun Fact: The house has 160 rooms, 2,000 doors, and 10,000 windows, with continuous construction for nearly 40 years.

22. The Mystery of the Monoliths

In late 2020, a series of mysterious metallic monoliths began appearing around the world. The first appeared in Utah, USA, and others followed in locations like Romania, California, and Turkey.

Fun Fact: Many people have speculated that the monoliths are a marketing stunt, a form of art, or even extraterrestrial in nature. Some monoliths were found to be art installations or pranks.

23. The Legend of the Fountain of Youth

The Fountain of Youth is a legendary spring that supposedly restores youth to anyone who drinks from it. The legend is most famously linked to Juan Ponce de León, a Spanish explorer who searched for it in the early 1500s in Florida.

Fun Fact: Despite being linked to Ponce de León, historians believe he may have never actually been searching for the Fountain of Youth.

24. The Enigma of the Taos Hum

The Taos Hum is a low-frequency sound heard by some residents and visitors of Taos, New Mexico. Despite numerous studies, the source of the hum remains a mystery. It's often described as a faint buzzing or droning sound, but not everyone can hear it.

Fun Fact: While only about 2% of the population can hear the hum, it has prompted scientific investigations to understand its origin.

25. The Secret of the Mokele-Mbembe

The Mokele-Mbembe is a creature reported to inhabit the Congo River Basin in Africa. Descriptions of the creature often resemble a sauropod dinosaur, suggesting that it may be a prehistoric survivor.

Fun Fact: Some believe the Mokele-Mbembe is simply a misidentified elephant or hippopotamus, while others insist it's a dinosaur-like creature.

26. The Nazca Lines and Ancient Astronomical Knowledge

The Nazca Lines have sparked theories that the Nazca people possessed advanced knowledge of astronomy. Some researchers believe the lines were constructed to mark celestial events or even used as ritual pathways to the gods.

Fun Fact: The largest Nazca geoglyph, depicting a hummingbird, measures over 300 feet in length.

27. The Unexplained Death of Napoleon Bonaparte

Napoleon Bonaparte, the famous French military leader, died in 1821 while in exile on the island of Saint Helena. The cause of his death has been debated for centuries. While many historians believe he died of stomach cancer, others suspect he was poisoned.

Fun Fact: The presence of arsenic in Napoleon's hair and nails has led to speculation that he may have been poisoned.

28. The Curse of the Pharaohs

The Curse of the Pharaohs is a legendary curse that supposedly affects those who disturb the tombs of ancient Egyptian rulers, especially Tutankhamun's tomb, which was discovered in 1922. Several members of the excavation team died mysteriously, fueling rumors that the curse was real.

Fun Fact: The idea of the Curse of the Pharaohs gained widespread attention after the death of Lord Carnarvon, the British aristocrat who funded the tomb's excavation.

29. The Enigma of the Pyramids of Giza

The Pyramids of Giza have stood for over 4,500 years, and they remain one of the most impressive architectural achievements in history. The pyramids were built to honor pharaohs, and their construction required thousands of workers and incredible engineering skills.

Fun Fact: The Great Pyramid was originally covered in smooth, white limestone, making it shine brightly in the sun. Most of this covering has been removed over time.

30. The Search for the Holy Grail

The Holy Grail, the legendary cup used by Jesus Christ at the Last Supper, has been a source of fascination and mystery for centuries. Countless expeditions, myths, and legends have surrounded the search for this sacred object.

Fun Fact: The Holy Grail is thought to possess mystical powers, including the ability to grant eternal life or healing.

31. The Legend of Lemuria

Lemuria is a hypothetical lost continent believed to have existed in the Indian and Pacific Oceans. The concept of Lemuria originated in the 19th century to explain the presence of similar fossils and geological formations in different parts of the world.

Fun Fact: While there is no scientific evidence to support its existence, Lemuria continues to fascinate and inspire myths and legends.

32. The Mystery of the Crystal Skulls

The Crystal Skulls are a collection of human skull-shaped artifacts made from quartz crystal. Found in Central America, these skulls are said to have mystical or supernatural properties.

Fun Fact: There are only a few known crystal skulls in existence, and most are considered modern hoaxes, but the legends surrounding them persist.

33. The Unexplained Mystery of the Marfa Lights

The Marfa Lights are mysterious glowing orbs that appear in the desert near Marfa, Texas. They've been observed for over a century, and their cause remains unknown.

Fun Fact: Explanations range from car headlights to atmospheric reflections, but no single theory fully explains the Marfa Lights.

34. The Phantom Time Hypothesis

The Phantom Time Hypothesis is a controversial theory suggesting that the Early Middle Ages (614-911 AD) never happened and were added to the calendar.

Fun Fact: While widely rejected by historians, the theory highlights the challenges of accurately reconstructing historical timelines.

35. The Mystery of the Pollock Twins

The Pollock Twins case involves two sisters who were believed to be the reincarnations of their deceased older sisters. The case is often cited as evidence for reincarnation.

Fun Fact: The surviving sisters reportedly exhibited knowledge and behaviors similar to their deceased siblings, despite having no prior exposure to that information.

36. The Lost Dutchman's Gold Mine

The Lost Dutchman's Gold Mine is a legendary mine said to be hidden somewhere in the Superstition Mountains of Arizona. Despite numerous searches, the mine has never been found.

Fun Fact: The mine is named after Jacob Waltz, a German immigrant ("Dutchman" was a common term for Germans at the time), who supposedly discovered it in the 19th century.

37. The Dyatlov Pass Incident

In 1959, nine experienced hikers died mysteriously in the Ural Mountains of Russia. The circumstances surrounding their deaths are bizarre and have led to numerous theories.

Fun Fact: The hikers' tent was found cut open from the inside, and some of the bodies had unexplained injuries, leading to speculation about paranormal activity or a secret military experiment.

38. The Bloop

"The Bloop" was an ultra-low-frequency, high-amplitude underwater sound detected by the U.S. National Oceanic and Atmospheric Administration (NOAA) in 1997. Its origin remains unknown.

Fun Fact: While some speculated that the sound was made by a giant sea creature, NOAA later suggested it was likely the sound of an icequake (an iceberg cracking).

39. Kaspar Hauser

Kaspar Hauser was a mysterious German youth who appeared in Nuremberg in 1828, claiming to have grown up in total isolation in a darkened cell. His origins and short life are shrouded in mystery.

Fun Fact: Hauser's story inspired numerous books, films, and plays, and he has been the subject of much speculation and debate.

40. The Oak Island Money Pit

The Oak Island Money Pit is a deep, elaborately constructed shaft on Oak Island in Nova Scotia, Canada. For centuries, people have believed that treasure is buried at the bottom of the pit.

Fun Fact: Despite numerous excavations, and even some fatalities, no significant treasure has ever been found, leading many to believe the "treasure" is a myth.

Conclusion:

The world is filled with **mysterious events** and **unsolved puzzles** that continue to spark our curiosity. Whether it's lost civilizations, strange disappearances, or cryptic artifacts, these enigmas remind us that there is still much to discover.

Though some of these mysteries may never be solved, they inspire us to keep asking questions and seeking answers. Who knows what hidden secrets are waiting to be uncovered?

Chapter 6:
Unexpected Natural Phenomena

Nature has a way of surprising us with its incredible, and sometimes baffling, displays.

In this chapter, we'll explore bizarre weather patterns, rare geological events, and extraordinary biological phenomena that continue to amaze scientists and challenge our understanding of the world.

Prepare to be astonished as we uncover the most **unexpected natural phenomena!**

1. Ball Lightning

A rare phenomenon where glowing orbs of light appear during thunderstorms, sometimes hovering or moving unpredictably before vanishing. These luminous spheres can range in size from a small marble to several feet in diameter and often last for a few seconds before disappearing with a loud pop or even an explosion.

Fun Fact: Some scientists believe ball lightning could be caused by vaporized silicon reacting with oxygen in the air, while others theorize it might be a form of plasma.

2. Raining Animals

Reports exist of fish, frogs, and even spiders falling from the sky—often due to waterspouts or strong winds lifting them into storm clouds before dropping them back to Earth. These bizarre downpours have been recorded throughout history, baffling those who witness them.

Fun Fact: In 2007, thousands of spiders "rained" over Salta, Argentina, covering the ground with webs! Some reports even suggest frozen frogs have fallen during winter storms.

3. The Sailing Stones of Death Valley

In the remote Racetrack Playa of Death Valley, heavy rocks mysteriously move across the desert floor, leaving long trails behind them. Scientists have linked the movement to thin ice sheets melting and pushing the rocks forward with the help of strong winds.

Fun Fact: Some stones have traveled over 1,500 feet (457 meters) without human or animal intervention! The phenomenon was first documented in the early 1900s but wasn't scientifically explained until 2014.

4. Bioluminescent Waves

Certain ocean waves glow at night due to bioluminescent plankton that light up when disturbed, creating a magical blue glow. This eerie yet beautiful sight has been observed in many coastal regions around the world.

Fun Fact: The beaches of the Maldives are famous for their stunning "Sea of Stars" effect caused by glowing plankton! Some species of bioluminescent plankton also react to predators by lighting up, potentially scaring them away.

5. The Everlasting Storm (Catatumbo Lightning)

A unique weather phenomenon in Venezuela where a lightning storm occurs almost every night over Lake Maracaibo. The constant electric activity is caused by humid air currents interacting with cold mountain breezes, creating an atmospheric storm factory.

Fun Fact: This storm can produce up to 280 lightning strikes per hour and is visible from over 250 miles away! It holds the Guinness World Record for the highest concentration of lightning per square kilometer.

6. Morning Glory Clouds

Extremely rare tube-shaped clouds that roll across the sky

like a long, continuous wave. These formations can stretch for hundreds of miles and appear suddenly, mesmerizing those who witness them.

Fun Fact: These clouds can stretch over 600 miles (1,000 km) and are most commonly seen in Northern Australia, particularly in Burketown, where pilots even ride the phenomenon for sport.

7. The Black Sun Murmurations

Hundreds of thousands of starlings form mesmerizing, shifting patterns in the sky, almost like a coordinated dance. These large flocks move in perfect harmony, twisting and turning as if controlled by a single mind.

Fun Fact: Scientists believe murmurations help birds avoid predators by confusing them with rapid, unpredictable movements. This phenomenon is most commonly seen in Denmark during the autumn migration.

8. Ice Circles

Large, perfectly circular slabs of ice that spin on the surface of slow-moving rivers, forming naturally in cold climates.

Fun Fact: Some ice circles can be over 50 feet (15 meters) wide and spin for hours without breaking apart.

9. Methane Bubbles Frozen in Lakes

Trapped methane gas from decomposing organic material

forms white bubbles beneath the ice, creating mesmerizing patterns.

Fun Fact: If these bubbles are popped and ignited, they can burst into flames!

10. Volcanic Lightning (Dirty Thunderstorms)

Lightning that occurs inside volcanic ash clouds due to friction between charged ash particles.

Fun Fact: This eerie phenomenon has been recorded in eruptions like Mount Vesuvius (79 AD) and Eyjafjallajökull (2010).

11. Blood Rain

This rare phenomenon occurs when rain appears red due to high concentrations of dust or algae spores in the atmosphere. Historical records show instances of blood rain dating back centuries, often regarded as omens.

Fun Fact: In 2001, Kerala, India, experienced blood rain for several weeks, which some scientists believe contained microorganisms from a possible meteor origin!

12. Frost Flowers

These delicate ice formations appear on plant stems or frozen surfaces when moisture freezes instantly in cold air. They look like delicate white petals, forming intricate frozen "blooms."

Fun Fact: Frost flowers are most commonly found in polar regions and along frozen lakes in early winter!

13. Fire Whirls (Fire Tornadoes)

A rare and terrifying phenomenon where intense heat and strong winds combine to create a tornado-like spinning column of fire. These can occur during wildfires, causing extreme destruction.

Fun Fact: In 1923, a fire whirl in Japan during the Great Kanto Earthquake reached speeds of 200 mph and killed thousands!

14. Blue Lava

Unlike traditional lava, some volcanoes release sulfuric gases that ignite in the air, producing an eerie **blue glow** instead of the typical red-orange lava. The best-known site for this phenomenon is the **Kawah Ijen** volcano in Indonesia.

Fun Fact: The blue color isn't the lava itself—it's the combustion of sulfuric gas at extremely high temperatures!

15. The Green Flash

This rare optical event happens at sunrise or sunset when a quick burst of green light appears just above the sun's edge. It occurs due to the Earth's atmosphere bending sunlight.

Fun Fact: Sailors once believed seeing the Green Flash meant they would gain wisdom or never be deceived again!

16. Brinicles ("Icicle of Death")

Brinicles are underwater icicles that form when super-cold brine sinks into the ocean, freezing water around it. These can trap and freeze marine life like starfish and sea urchins.

Fun Fact: Brinicles were first filmed in Antarctica in 2011, revealing their eerie slow-motion freezing process!

17. Magnetic Hill

An optical illusion where a road appears to slope uphill, but objects (like cars) roll backward as if pulled by an invisible force. These "gravity hills" exist in many locations around the world.

Fun Fact: Scientists believe Magnetic Hills are caused by tilted horizon lines that trick the human brain into perceiving slopes incorrectly!

18. The Aurora Borealis (Northern Lights)

One of nature's most breathtaking spectacles, these colorful lights appear in the sky due to charged particles from the sun colliding with Earth's magnetic field.

Fun Fact: There is also a Southern Hemisphere version called the **Aurora Australis** (Southern Lights)!

19. The Devil's Kettle Waterfall

Located in Minnesota, this mysterious waterfall splits into two streams—one side continues downstream, but the other vanishes into a deep hole, with no one knowing exactly where it ends!

Fun Fact: Despite extensive dye and GPS tracking, scientists still haven't fully solved where the lost water goes!

20. The Deepest Blue Hole (Dragon Hole, China)

This underwater sinkhole in the South China Sea is over **300 meters (987 feet) deep**, making it the deepest blue hole on Earth.

Fun Fact: Ancient legends say Dragon Hole is the mythical home of the Monkey King from Chinese folklore!

Fun Fact: The Ford Model T was nicknamed the "Tin Lizzie."

21. The Sailing Stones of Death Valley

These mysterious rocks move across the desert floor on their own, leaving long trails behind them. Scientists discovered that thin layers of ice, combined with strong winds, push the stones.

Fun Fact: Some rocks have traveled over 1,500 feet without human or animal interference!

22. The Boiling River of the Amazon

Deep in the Amazon rainforest, there is a river so hot that it boils small animals alive. The water reaches temperatures of **200°F (93°C)**, and scientists believe underground geothermal activity causes the extreme heat.

Fun Fact: Local legends say the river was created by a giant serpent spirit!

23. The Door to Hell (Darvaza Gas Crater, Turkmenistan)

This fiery pit has been burning continuously since **1971**, when geologists accidentally ignited natural gas reserves underground.

Fun Fact: Despite the danger, tourists visit this glowing crater, which measures **230 feet wide and 98 feet deep**!

24. The Crooked Forest (Poland)

A collection of over 400 pine trees in Poland grows with bizarre **J-shaped bends at the base**, forming a hauntingly symmetrical landscape.

Fun Fact: No one knows exactly why these trees grow this way—some believe it's due to human intervention, while others think extreme weather played a role.

25. The Living Rocks of Chile (Pyura chilensis)

These strange sea creatures look like **ordinary rocks** but are actually alive! They filter seawater for nutrients and contain bright red blood filled with vanadium, a rare metal.

Fun Fact: Locals harvest these "rocks" as seafood and claim they taste like iodine!

26. The Underwater Waterfall of Mauritius

Off the coast of Mauritius, an incredible optical illusion makes it appear as if a **massive waterfall is plunging into the ocean depths**—but it's actually sand being pulled down by ocean currents.

Fun Fact: This natural illusion is best seen from above, making it a popular spot for aerial photography!

27. The Fairy Circles of Namibia

Mysterious circular patches of **barren land** dot the grasslands of Namibia, and scientists still debate whether termites or underground gases are responsible for these formations.

Fun Fact: These circles can remain unchanged for decades, giving the landscape an otherworldly appearance.

28. The Pink Lake (Lake Hillier, Australia)

Unlike most lakes, Lake Hillier in Australia stays a **bright pink** year-round due to salt-loving algae and bacteria that produce red pigments.

Fun Fact: Even when removed from the lake, the water remains pink in a bottle!

29. The Coldest Place on Earth (Antarctica's Ridge A)

This remote location in Antarctica holds the record for the **coldest temperature ever recorded: −128.6°F (−89.2°C)**.

Fun Fact: Scientists say that if you breathed in without protection here, your lungs would freeze instantly!

30. The Humming Sand Dunes

In certain deserts, sand dunes produce deep, **humming sounds** when disturbed. This eerie noise is caused by layers of sand vibrating together.

Fun Fact: Marco Polo described this phenomenon in his travels, thinking it was caused by evil spirits!

31. The Vanishing Island (Bermeja, Mexico)

An island that appeared on maps for centuries suddenly **vanished without a trace**! Some believe it was swallowed by rising sea levels, while others think it never existed.

Fun Fact: In 2009, Mexican officials sent a team to locate Bermeja—but they found nothing but open ocean!

32. The Giant's Causeway (Ireland)

A natural wonder made up of **40,000 hexagonal basalt columns**, formed by ancient volcanic activity. Legends say it was built by giants trying to cross the sea to Scotland.

Fun Fact: The columns fit together so perfectly that they look man-made!

33. The Immortal Jellyfish (Turritopsis dohrnii)

This tiny jellyfish has the incredible ability to **reverse its aging process**, making it biologically immortal. When injured or stressed, it transforms back into a juvenile state and starts its life cycle again.

Fun Fact: Scientists are studying this jellyfish to understand aging and potential applications for human medicine!

34. The Blood Falls of Antarctica

A blood-red waterfall flows from a glacier in Antarctica, colored by iron-rich water that oxidizes when exposed to air.

Fun Fact: This eerie sight was once thought to be caused by algae, but it was later proven to be mineral-rich water trapped underground for millions of years.

35. The Phantom Island of Sandy Island

An island **appeared on maps for over 100 years**, but when researchers visited the site in 2012, they found nothing but **empty ocean**!

Fun Fact: Some believe early explorers misinterpreted volcanic pumice floating on the water as land.

36. The Devil's Pool (Victoria Falls, Africa)

This natural rock pool sits at the very **edge of Victoria Falls**, one of the world's largest waterfalls. During the dry season, adventurous swimmers can bathe just inches from the deadly drop!

Fun Fact: Despite its terrifying location, the pool is safe when water levels are low!

37. The Eternal Flame Falls (New York, USA)

A small waterfall hides a **naturally burning flame** behind it, fueled by underground gas leaks. The combination of water and fire creates a mystical sight.

Fun Fact: If the flame ever goes out, hikers can relight it with a match!

38. The Petrifying Well (England)

A rare well that turns objects to **stone** over time by coating them in mineral deposits. Items left in the water gradually become rock-like, as if cursed.

Fun Fact: Locals once believed this well was haunted by witches due to its eerie powers!

39. The Red Crab Migration (Christmas Island, Australia)

Every year, millions of bright red crabs migrate across the island, covering roads, beaches, and forests in a **moving red sea** as they head to the ocean to lay their eggs.

Fun Fact: To protect the crabs, roads are shut down, and special bridges have been built for them to cross!

40. The Ice Caves of Iceland

During winter, Iceland's glaciers reveal hidden **crystal-blue ice caves**, creating a breathtaking underground world. These caves constantly shift and melt, meaning no two seasons look the same.

Fun Fact: The ice can be thousands of years old, trapping ancient air bubbles inside!

Conclusion:

Nature is full of wonders that defy expectations! From glowing waves to wandering stones, these extraordinary phenomena remind us that there is still so much to learn about our planet. Every new discovery challenges our understanding and pushes the boundaries of science. Keep questioning, keep exploring, and who knows? You might just witness one of these mysteries with your own eyes!.

Chapter 7:
History's Greatest Conspiracies

Throughout history, there have been events so shrouded in secrecy, deception, and unanswered questions that they have given rise to some of the world's most enduring conspiracies.

Were secret societies influencing world events? Did lost civilizations leave behind hidden knowledge?

In this chapter, we'll uncover **history's greatest conspiracies**—the unsolved puzzles and whispered theories that continue to spark debate. Keep your mind open, and let's explore these historical enigmas!

1. The Disappearance of the Ninth Roman Legion

The elite Ninth Legion of Rome mysteriously vanished from records after being sent to Britain. Did they fall in battle, or was their fate covered up by the Empire?

Fun Fact: Some theories suggest they were ambushed in Scotland, while others believe they defected and started a new civilization elsewhere.

2. The Secret of the Vatican Archives

The Vatican holds an enormous collection of historical documents, some of which have been kept from the public for centuries. What hidden knowledge lies within?

Fun Fact: The archives contain letters from Michelangelo, documents on the trial of the Knights Templar, and even correspondence with Adolf Hitler.

3. The Lost Treasure of the Knights Templar

The legendary Knights Templar were rumored to have amassed a fortune, including the Holy Grail and other sacred relics, before disappearing in the 14th century.

Fun Fact: Some believe the treasure is hidden beneath Rosslyn Chapel in Scotland, while others think it was taken to North America.

4. The Mystery of Oak Island's Money Pit

A mysterious pit on Oak Island, Canada, has been the subject of treasure hunts for over 200 years. Some believe it contains pirate gold, while others say it hides lost artifacts from Europe.

Fun Fact: Even Franklin D. Roosevelt invested in expeditions to uncover the island's secrets.

5. The Man in the Iron Mask

A prisoner was held in France's Bastille for decades, forced to wear a mask to conceal his identity. Was he a disgraced noble, the twin brother of King Louis XIV, or someone even more dangerous?

Fun Fact: His identity remains unknown, but some suspect he was a political threat to the monarchy.

6. The True Fate of Anastasia Romanov

The youngest daughter of Russia's last czar was believed to have been executed with her family in 1918, yet rumors persisted for decades that she had survived and gone into hiding.

Fun Fact: In 2007, DNA testing confirmed that Anastasia and her family had indeed perished, finally ending the mystery.

7. The Death of Napoleon Bonaparte

Official records state that Napoleon died of stomach cancer in exile, but traces of arsenic found in his hair led some to believe he was poisoned.

Fun Fact: Some historians argue that arsenic levels in Napoleon's body were due to environmental exposure from the wallpaper in his home.

8. The Identity of Jack the Ripper

One of history's most infamous serial killers terrorized London in 1888, yet his identity remains a mystery. Numerous suspects have been proposed, from doctors to members of the royal family.

Fun Fact: Some modern DNA analysis claims to have identified Jack the Ripper as a Polish barber named Aaron Kosminski, though this remains debated.

9. The Hidden Chamber in the Great Pyramid

Some Egyptologists believe there is a hidden chamber within the Great Pyramid of Giza, possibly containing undiscovered artifacts or even the lost knowledge of ancient civilizations.

Fun Fact: In 2017, scientists using cosmic rays detected a large void inside the pyramid, but its purpose remains unknown.

10. The Lost City of Atlantis

Plato described a powerful and advanced civilization that was swallowed by the sea. Was Atlantis a real place, or merely an allegory for human hubris?

Fun Fact: Some believe Atlantis may have been inspired by the Minoan civilization, which was devastated by a volcanic eruption on the island of Thera (modern-day Santorini).

11. The Philadelphia Experiment

A secret U.S. Navy experiment in 1943 allegedly made the USS Eldridge invisible and teleported it to another location. Some claim the event resulted in bizarre side effects, including crew members merging with the ship itself.

Fun Fact: The U.S. Navy denies the experiment ever happened, but conspiracy theorists insist there was a cover-up.

12. The Tunguska Explosion

In 1908, a massive explosion flattened over 800 square miles of Siberian forest. The cause is believed to be a meteor airburst, yet no impact crater was ever found.

Fun Fact: Some alternative theories suggest the explosion was caused by a comet, a volcanic gas eruption, or even an alien spacecraft.

13. The Die Glocke (The Nazi Bell)

Rumors persist that Nazi Germany developed an advanced, bell-shaped anti-gravity device during World War II. Some claim it was a secret weapon, while others believe it was an experiment in time travel.

Fun Fact: No physical evidence of Die Glocke has ever been found, making it one of the war's greatest mysteries.

14. The Secret Society of the Illuminati

Founded in 1776, the Illuminati was a real organization that sought to influence political and religious thought. Some believe they still exist today, manipulating world events from behind the scenes.

Fun Fact: The Illuminati were officially disbanded in the late 1700s, but conspiracy theories about their continued influence remain popular.

15. The Mystery of DB Cooper

In 1971, an unknown man hijacked a plane, demanded ransom money, then parachuted into the wilderness—never to be seen again. Despite decades of investigation, his identity and fate remain unknown.

Fun Fact: Some of the stolen money was found buried along the Columbia River in 1980, but Cooper himself was never located.

Fun Fact: Fawcett believed the city held vast treasures.

16. The Missing Amber Room

A magnificent chamber made entirely of amber panels, gold leaf, and mirrors, the Amber Room was stolen by Nazi forces during World War II and disappeared without a trace. Some believe it was destroyed, while others think it remains hidden.

Fun Fact: In 2003, a full reconstruction of the Amber Room was completed in Russia, using historical records and old photographs.

17. The Voynich Manuscript

This mysterious book, written in an unknown language with strange illustrations of plants and celestial diagrams, has baffled cryptographers for centuries. No one has been able to decode its meaning.

Fun Fact: Some theories suggest it is a medical or alchemical text, while others believe it is an elaborate hoax.

18. The Secret Gold of the Confederate Treasury

At the end of the American Civil War, the Confederate treasury vanished. Some believe the gold was smuggled away by Southern loyalists, while others think it was hidden to fund a future rebellion.

Fun Fact: Treasure hunters still search for the missing Confederate gold, with rumors pointing to locations in Georgia, Virginia, and Michigan.

19. The Black Dahlia Murder

The gruesome 1947 murder of aspiring actress Elizabeth Short remains one of Hollywood's most famous unsolved cases.

Many suspects were investigated, but the identity of her killer was never confirmed.

Fun Fact: Some theories link the murder to a surgeon due to the precise nature of the wounds, but no definitive proof has been found.

20. The Great Train Robbery of 1963

A group of criminals successfully stole over £2.6 million from a Royal Mail train in England, using minimal violence and careful planning. While most of the gang was caught, a significant portion of the money was never recovered.

Fun Fact: One of the masterminds, Ronnie Biggs, escaped prison and lived in Brazil for decades before voluntarily returning to the UK.

21. The Green Children of Woolpit

Medieval accounts tell of two children with green skin who appeared in an English village, speaking an unknown language. Some believe they were lost travelers, while others think they were from a parallel world.

Fun Fact: The children eventually adapted to normal food, and their skin color faded—but the mystery of their origins remains unsolved.

22. The Shroud of Turin

Believed by some to be the burial cloth of Jesus, the Shroud of Turin bears the faint image of a man who appears to have suffered crucifixion wounds. Scientists are divided on whether it is an authentic relic or a medieval forgery.

Fun Fact: Carbon dating tests suggest the shroud dates to the Middle Ages, but some argue the tests were flawed due to contamination.

23. The Hidden Treasure of Forrest Fenn

In 2010, art dealer Forrest Fenn claimed to have hidden a chest filled with gold and jewels somewhere in the Rocky Mountains, sparking a decade-long treasure hunt.

Fun Fact: The treasure was finally discovered in 2020, but Fenn never revealed the exact location before his death.

24. The Dyatlov Pass Incident

In 1959, nine hikers mysteriously died in the Ural Mountains under bizarre circumstances, with some found half-dressed, others suffering severe internal injuries, and traces of radiation on their clothing.

Fun Fact: Theories range from an avalanche to secret military experiments, but the case remains unresolved.

25. The Death of Meriwether Lewis

One of the leaders of the Lewis and Clark Expedition, Meriwether Lewis, was found dead in 1809. Officially ruled a suicide, some believe he was assassinated due to political tensions or secrets he uncovered.

Fun Fact: His family and some historians have called for an official exhumation to determine if he was murdered.

26. The Montauk Project

A supposed secret government experiment at Montauk, New York, allegedly involved mind control, time travel, and contact with extraterrestrials. Conspiracy theorists claim the project was real and inspired the show *Stranger Things*.

Fun Fact: No concrete evidence of the Montauk Project exists, but many claim they were involved in experiments there.

27. The Lost Dutchman's Gold Mine

A legendary gold mine hidden in the Superstition Mountains of Arizona has lured treasure hunters for centuries. Despite numerous expeditions, no one has ever found the mine.

Fun Fact: Some claim the mine is cursed, and many who have searched for it have vanished or died under mysterious circumstances.

28. The Skeleton Lake of Roopkund

This remote Himalayan lake is filled with hundreds of ancient skeletons, believed to be over 1,000 years old. The cause of their deaths is still debated, with theories ranging from an epidemic to a sudden hailstorm.

Fun Fact: Genetic testing showed the skeletons belonged to people from different regions, suggesting a diverse group perished there.

29. The Tungsten Mystery of World War II

During World War II, large amounts of tungsten—a rare metal used for weapons—were smuggled through Spain. Who was behind the smuggling, and where did all the profits go?

Fun Fact: Some researchers believe both Allied and Axis powers had secret deals regarding tungsten supply chains.

30. The Ghost Ship Mary Celeste

In 1872, the Mary Celeste was found adrift in the Atlantic Ocean with its crew missing, yet no signs of struggle or damage. The fate of the crew remains a mystery.

Fun Fact: The ship's cargo and supplies were untouched, deepening the mystery of what caused the crew to abandon ship.

31. The Lost Cosmonauts Theory

Some believe the Soviet Union sent cosmonauts into space before Yuri Gagarin's historic flight but covered up their deaths when the missions failed.

Fun Fact: Soviet records deny this claim, but leaked audio recordings allegedly capture distress signals from lost astronauts.

32. The Secrets of Area 51

Long associated with UFOs and secret military experiments, Area 51 has fueled speculation for decades. While the U.S. government acknowledges its existence, its true purpose remains classified.

Fun Fact: Declassified documents reveal Area 51 was used to test spy planes during the Cold War.

33. The Curse of the Hope Diamond

This famous blue diamond is said to bring misfortune to its owners. Many who possessed it reportedly faced tragedy or death.

Fun Fact: The diamond is now on display at the Smithsonian Institution, and no new misfortunes have been reported.

34. The Nazi Gold Train

Legends claim a Nazi train loaded with gold disappeared in Poland near the end of World War II. Many have searched for it, but it has never been found.

Fun Fact: In 2015, two treasure hunters claimed to have located the train, but excavation efforts were inconclusive.

35. The Taman Shud Case

An unidentified man was found dead on an Australian beach

in 1948 with a cryptic note in his pocket that read "Taman Shud," meaning "ended" in Persian. His identity and cause of death remain unknown.

Fun Fact: The note was linked to a rare book of poetry, but how it relates to the case is still a mystery.

36. The Lascaux Cave Paintings

These ancient cave paintings in France contain stunning images of animals and mysterious symbols. Some believe they hold lost knowledge or were used for spiritual rituals.

Fun Fact: The caves were closed to the public after the paintings began deteriorating due to human exposure.

37. The Mystery of Rennes-le-Château

A small French village became famous after rumors spread that a priest discovered vast hidden wealth and used it to renovate the local church.

Fun Fact: Some conspiracy theorists link the treasure to the Holy Grail or lost Templar riches.

38. The Zodiac Killer's Ciphers

The infamous Zodiac Killer sent cryptic messages to newspapers in the 1960s, some of which remained unsolved for decades.

Fun Fact: In 2020, a team of codebreakers finally solved one of the Zodiac's ciphers, but his identity is still unknown.

39. The Oak Ridge Secret City

During World War II, an entire city in Tennessee was built in secret to support the Manhattan Project. Workers had no idea they were helping create the atomic bomb.

Fun Fact: The city wasn't listed on any map, and security was so tight that even employees didn't fully understand what they were working on.

40. The Lost Colony of Roanoke

An entire English colony disappeared in the 16th century, leaving behind only the word "CROATOAN" carved into a tree.

Fun Fact: The fate of the settlers remains unknown, with theories ranging from assimilation with Native Americans to massacre.

Conclusion

History is filled with mysteries, but perhaps the greatest mystery is how much truth lies beneath the surface. Some conspiracies may be dismissed as myths, while others contain fragments of reality waiting to be uncovered. Keep questioning the past, and you might just rewrite history yourself!

Chapter 8:
Breakthroughs That Changed the World

Throughout history, scientific discoveries and technological innovations have reshaped our understanding of the world.

From the development of medicine to the exploration of space, these breakthroughs have not only advanced civilization but also changed the course of human history.

In this chapter, we will explore some of the most significant discoveries and inventions that have revolutionized science, technology, and our way of life.

1. The Discovery of CRISPR Gene Editing

CRISPR is a groundbreaking technology that allows scientists to edit DNA with precision, opening doors for genetic disease treatment, improved agriculture, and even potential cures for inherited disorders.

Fun Fact: CRISPR was inspired by bacteria that use a natural gene-editing process to defend against viruses.

2. The Development of mRNA Vaccines

mRNA vaccine technology revolutionized immunization by using genetic instructions to trigger an immune response. The COVID-19 vaccines were the first widely used mRNA vaccines, demonstrating their effectiveness.

Fun Fact: mRNA vaccine research had been ongoing for decades, but the pandemic accelerated its real-world application.

3. The Invention of X-Rays

Wilhelm Roentgen accidentally discovered X-rays in 1895 while experimenting with cathode rays. This breakthrough allowed doctors to see inside the human body without surgery.

Fun Fact: The first X-ray image was of Roentgen's wife's hand, revealing her bones and wedding ring.

4. The First Image of a Black Hole

In 2019, a global network of telescopes captured the first-ever image of a black hole, confirming long-standing predictions in physics.

Fun Fact: The black hole is located in the M87 galaxy and has a mass **6.5 billion times** that of our Sun.

5. The Discovery of Neutrinos

Neutrinos are nearly massless subatomic particles that pass through matter unnoticed. Studying them has helped scientists understand fundamental physics.

Fun Fact: Trillions of neutrinos pass through your body every second, but they rarely interact with anything.

6. The Expansion of the Universe

Astronomer Edwin Hubble discovered that galaxies are moving away from each other, proving that the universe is expanding and leading to the Big Bang theory.

Fun Fact: The farther a galaxy is, the faster it moves away—this discovery changed our understanding of the cosmos.

7. The Discovery of Dark Matter and Dark Energy

Scientists found that **95% of the universe is made of unknown substances**—dark matter and dark energy—but their exact nature remains a mystery.

Fun Fact: Dark energy is believed to be the force behind the accelerating expansion of the universe.

8. The Invention of the Microchip

Microchips revolutionized computing by allowing electronic devices to become smaller and faster. This led to the development of modern computers, smartphones, and nearly all digital technology.

Fun Fact: The first microchip had only a few transistors, while today's chips have **billions** of them.

9. The Discovery of the Ozone Hole

Scientists found that chemicals used in aerosols and refrigerants were destroying the ozone layer, leading to international agreements that helped slow the damage.

Fun Fact: The ozone layer is slowly recovering thanks to environmental protections.

10. The Confirmation of Gravitational Waves

Albert Einstein predicted gravitational waves, but they weren't detected until 2015, proving that massive objects like black holes can create ripples in space-time.

Fun Fact: The waves detected in 2015 were caused by two black holes colliding over **a billion years ago.**

11. The Invention of the Laser

Lasers transformed medicine, communications, and technology, leading to barcode scanners, laser surgery, and fiber-optic internet.

Fun Fact: The word "laser" stands for **Light Amplification by Stimulated Emission of Radiation.**

12. The Discovery of Extremophiles

Scientists found that life thrives in extreme environments like deep-sea vents and Antarctica, challenging our understanding of where life can exist.

Fun Fact: Some extremophiles could **potentially survive on Mars.**

13. The Development of Artificial Intelligence (AI)

AI has evolved from basic computer programs to complex machine-learning models that can recognize speech, generate art, and even drive cars.

Fun Fact: The first AI program was created in 1951 and played chess. Today, AI helps power search engines, medical diagnoses, and space exploration.

14. The Discovery of the Higgs Boson ("God Particle")

The Higgs boson is a subatomic particle that gives other particles mass, confirming a major theory in physics.

Fun Fact: It was discovered in 2012 at the Large Hadron Collider, one of the most powerful scientific machines ever built.

15. The Discovery of Exoplanets

In 1992, astronomers confirmed the existence of planets beyond our solar system, reshaping our understanding of planetary systems. Thousands of exoplanets have since been discovered, some potentially habitable.

Fun Fact: The first exoplanets found were orbiting a pulsar, rather than a Sun-like star.

16. The Discovery of the Cosmic Microwave Background Radiation

In 1964, scientists detected faint radiation across the universe, providing crucial evidence for the Big Bang theory.

Fun Fact: This discovery was made by accident when radio astronomers detected an unexplained background noise that turned out to be leftover radiation from the early universe.

17. The Invention of GPS Technology

The Global Positioning System (GPS) was developed for military navigation but has since transformed travel, communication, and daily life worldwide.

Fun Fact: GPS signals travel at the speed of light and must account for Einstein's relativity to ensure accuracy.

18. The Discovery of Water on Mars

Multiple NASA missions have confirmed the presence of frozen water and even seasonal liquid water on Mars, raising hopes for potential life.

Fun Fact: Some Mars craters contain ice that could be a future water source for astronauts.

19. The First Lab-Grown Organ Transplant

Scientists successfully grew and transplanted artificial organs using stem cells, marking a major step toward regenerative medicine.

Fun Fact: The first lab-grown organ successfully transplanted was a **bladder** in 1999.

20. The Creation of the First Synthetic Life Form

In 2010, scientists engineered the first living organism with an entirely synthetic genome, demonstrating the potential of genetic design.

Fun Fact: The synthetic bacteria were nicknamed **"Synthia"** and could be programmed for specific tasks, such as producing biofuels.

21. The Discovery of Superconductors

Materials that conduct electricity without resistance, superconductors have revolutionized MRI machines, maglev trains, and quantum computing.

Fun Fact: Some superconductors work at relatively high temperatures, but most require extreme cooling.

22. The First Human Spaceflight

In 1961, Yuri Gagarin became the first human in space, orbiting Earth aboard the Soviet spacecraft Vostok 1.

Fun Fact: Gagarin's famous phrase before launch was "Poyekhali!" which means **"Let's go!"** in Russian.

23. The Invention of the World Wide Web

Tim Berners-Lee created the first web browser in 1989, laying the foundation for the internet revolution.

Fun Fact: The world's first website, created by Berners-Lee, is still online today.

24. The Invention of Wireless Electricity

Scientists have developed ways to transmit electricity without wires, inspired by Nikola Tesla's early experiments.

Fun Fact: Wireless charging is now used in smartphones, electric vehicles, and medical implants.

25. The Mapping of the Human Microbiome

Scientists have identified trillions of bacteria, viruses, and fungi living in and on the human body, which play a crucial role in health.

Fun Fact: Your microbiome can be as unique as your fingerprint!

26. The First Successful Face Transplant

In 2005, doctors performed the first partial face transplant, a major breakthrough in reconstructive surgery.

Fun Fact: Full face transplants are now possible, helping patients with severe facial injuries regain function.

27. The Discovery of a Possible Ninth Planet

Scientists have found indirect evidence suggesting a giant, undiscovered planet lurking in the outer solar system.

Fun Fact: This hypothetical "Planet Nine" could be **10 times the mass of Earth** and take thousands of years to orbit the Sun.

28. The Invention of Reusable Rockets

SpaceX revolutionized space travel by developing rockets that can land and be reused, drastically reducing launch costs.

Fun Fact: The Falcon 9 rocket has completed multiple missions using the same booster.

29. The Development of Self-Driving Cars

AI-powered autonomous vehicles are being tested around the world, with potential to revolutionize transportation.

Fun Fact: The first self-driving car prototypes were created in the 1980s.

30. The Discovery of the First Interstellar Object

In 2017, astronomers detected 'Oumuamua, a cigar-shaped rock that came from another star system, making it the first confirmed interstellar object.

Fun Fact: Some scientists speculate it could be an artificial probe, though no evidence supports this theory.

31. The Creation of Artificial Leaves

Scientists have developed synthetic leaves that can mimic photosynthesis, turning sunlight into fuel.

Fun Fact: These artificial leaves could help reduce carbon emissions by converting CO_2 into useful energy.

32. The Discovery of Gravitational Lensing

Einstein predicted that massive objects can bend light, and astronomers have used this effect to study distant galaxies.

Fun Fact: Some gravitational lensing events create "Einstein rings," where a galaxy appears as a glowing circle around another object.

33. The First Space Hotel Plans

Private companies are planning to build **hotels in space**, where tourists could stay in Earth's orbit.

Fun Fact: The first space hotel prototype may launch by 2030.

34. The Advancement of Cloning Technology

From Dolly the Sheep to potential human applications, cloning continues to be an area of ethical and scientific exploration.

Fun Fact: Scientists have cloned everything from cows to endangered species.

35. The Creation of Biodegradable Plastic

New materials made from plants and bacteria are helping to fight pollution by breaking down naturally.

Fun Fact: Some biodegradable plastics can dissolve in water without harming the environment.

36. The Invention of 3D-Printed Organs

Scientists are working on printing human tissues and even entire organs using 3D printers.

Fun Fact: The first 3D-printed heart prototype was created using human cells.

37. The First Private Mission to the Moon

Private companies are competing to send commercial landers and even tourists to the Moon.

Fun Fact: NASA has partnered with companies like SpaceX to make lunar tourism a reality.

38. The Discovery of an Earth-Like Exoplanet

Astronomers have found planets in the habitable zones of distant stars, raising hopes for finding extraterrestrial life.

Fun Fact: The most promising exoplanet, Proxima Centauri b, is only **4.2 light-years away**.

39. The Advancement of Brain-Computer Interfaces

Neural implants could one day allow people to control devices with their thoughts, helping those with paralysis communicate and move.

Fun Fact: Some brain implants can already help restore partial vision.

40. The Next Generation of Nuclear Fusion

Scientists are working toward clean, limitless energy using nuclear fusion, the process that powers the Sun.

Fun Fact: Fusion reactors could produce **millions of times more energy than coal** without dangerous waste.

Conclusion:

Scientific discoveries are the building blocks of human progress. From unlocking the secrets of the universe to developing technologies that improve everyday life, these breakthroughs shape the world we live in. What was once science fiction is now reality, and the future holds even greater possibilities. The next life-changing discovery could come from an unexpected experiment or a curious mind asking the right question.

Keep exploring, keep questioning, and who knows? The next great discovery might just come from you.

Chapter 9:
Unbelievable Feats of Human Ingenuity

Throughout history, humans have achieved extraordinary things, overcoming obstacles and pushing the boundaries of what was once thought impossible.

From **engineering wonders** and **groundbreaking inventions** to **unbelievable expeditions** and **pioneering efforts**, this chapter will highlight some of the most impressive feats of human ingenuity.

These accomplishments showcase the creativity, resilience, and determination that have shaped the world we live in today.

1. The Construction of the Great Pyramids

The Great Pyramids of Giza, one of the Seven Wonders of the Ancient World, were constructed around 4,500 years ago. These massive stone structures, built to honor the pharaohs, are considered one of the greatest engineering feats in history. The precision required to build these monuments with such large stones is still debated today.

Fun Fact: The Great Pyramid of Giza was the tallest man-made structure on Earth for over 3,800 years.

---

2. The Moon Landing

On July 20, 1969, Neil Armstrong became the first human to set foot on the Moon, marking one of humanity's greatest achievements. The Apollo 11 mission, led by NASA, successfully landed two astronauts on the lunar surface and safely returned them to Earth. This incredible feat was a result of years of scientific research, technological development, and international collaboration.

Fun Fact: Neil Armstrong's famous words when stepping onto the Moon were: "That's one small step for [a] man, one giant leap for mankind."

---

3. The Construction of the Eiffel Tower

The Eiffel Tower, completed in 1889, was initially met with criticism but later became one of the world's most famous landmarks. Designed by Gustave Eiffel, it was an innovative feat of iron construction and stood as the tallest man-made structure in the world until the completion of the Chrysler Building in New York in 1930.

Fun Fact: The Eiffel Tower was originally meant to be temporary, as it was built for the 1889 World's Fair, but it remains one of the most visited monuments in the world today.

---

4. The Building of the Great Wall of China

The Great Wall of China is a series of fortifications built by various Chinese dynasties over several centuries. Stretching over 13,000 miles, it was built to protect China from invasions and raids by nomadic tribes. The construction of this vast wall required the labor of millions of workers, including soldiers, peasants, and prisoners.

Fun Fact: The Great Wall is not a single continuous structure, but rather a series of walls and fortifications built at different times in history.

---

5. The Development of the Internet

The internet, which connects billions of people worldwide, is one of the most significant inventions of the modern era. Initially developed by the U.S. Department of Defense in the 1960s as ARPANET, the internet has revolutionized communication, commerce, and entertainment, shaping the way we live and work today.

Fun Fact: The first website was launched in 1991 by Tim Berners-Lee and is still accessible online as a piece of internet history.

---

6. The Creation of the Panama Canal

The Panama Canal is a man-made waterway that connects the Atlantic and Pacific Oceans, allowing ships to bypass the long and dangerous journey around the southern tip of South America. The construction of the canal was an incredible engineering challenge, requiring the movement of millions of cubic yards of earth and overcoming significant health risks from diseases like malaria and yellow fever.

Fun Fact: The Panama Canal was completed in 1914, and its construction is considered one of the greatest engineering feats of the 20th century.

---

7. The Discovery of the Polio Vaccine

In 1955, Jonas Salk developed the first effective vaccine for polio, a disease that caused paralysis and death worldwide. The vaccine's development was a major medical breakthrough and has helped virtually eliminate polio as a global health threat.

Fun Fact: Jonas Salk famously chose not to patent the polio vaccine, allowing it to be distributed worldwide at a low cost.

---

8. The Creation of the Suez Canal

The Suez Canal, completed in 1869, is an artificial waterway connecting the Mediterranean Sea and the Red Sea. This vital shipping route has played a key role in global trade by providing a shortcut for ships traveling between Europe and Asia.

Fun Fact: The Suez Canal was one of the most significant engineering projects of its time and remains one of the busiest maritime passages in the world.

9. Reaching for the Sky: Building the First Skyscraper

The construction of the first skyscrapers in the late 19th century, particularly the Home Insurance Building in Chicago, was a revolutionary feat of engineering. These towering structures, made possible by innovations in steel-frame construction and elevators, redefined urban skylines and transformed city living.

Fun Fact: The Home Insurance Building, completed in 1885, is often considered the first true skyscraper, although it stood only 10 stories tall.

---

10. The First Human Heart Transplant

In 1967, Dr. Christiaan Barnard performed the first successful human heart transplant in South Africa. The patient, Louis Washkansky, lived for 18 days after the transplant. While heart transplantation has improved significantly, this first surgery set the stage for organ transplants to become a life-saving medical procedure.

Fun Fact: Today, heart transplant recipients can live many years after their surgery, thanks to advancements in medical care and immunosuppressive drugs.

---

11. The Creation of the Golden Gate Bridge

Completed in 1937, the Golden Gate Bridge in San Francisco is considered one of the most beautiful and recognizable bridges in the world. Its construction involved overcoming challenges such as strong currents, frequent fog, and earthquakes, and it remains an iconic symbol of engineering ingenuity.

Fun Fact: The Golden Gate Bridge was the longest suspension bridge in the world when it was completed.

12. The Launch of the Hubble Space Telescope

In 1990, the Hubble Space Telescope was launched into orbit around Earth, providing astronomers with unprecedented views of distant galaxies, nebulae, and other cosmic phenomena. Hubble has transformed our understanding of the universe, contributing to major discoveries about black holes, dark matter, and the expansion of the universe.

Fun Fact: The Hubble Space Telescope has taken over 1 million observations of space, greatly enhancing our knowledge of the cosmos.

13. The Creation of the International Space Station (ISS)

The International Space Station (ISS), a joint project between NASA, Roscosmos, and other space agencies, is the largest space station ever built. Since its launch in 1998, astronauts from around the world have lived and worked aboard the ISS, conducting scientific research in microgravity.

Fun Fact: The ISS orbits Earth at an average altitude of about 400 kilometers and travels at a speed of 28,000 kilometers per hour.

14. The Completion of the Burj Khalifa

The Burj Khalifa in Dubai, completed in 2010, is the tallest building in the world, standing at 828 meters (2,717 feet). Its construction required cutting-edge technology and engineering solutions to support its immense height, and it stands as a testament to modern architectural and engineering capabilities.

Fun Fact: The Burj Khalifa has more than 160 floors and is home to luxury apartments, office spaces, and a hotel.

15. The First Successful Climb of Mount Everest

In 1953, Sir Edmund Hillary of New Zealand and Tenzing Norgay, a Sherpa of Nepal, became the first climbers to reach the summit of Mount Everest, the world's tallest mountain. Their achievement was a remarkable feat of endurance, skill, and determination, and it remains one of the greatest accomplishments in mountaineering.

Fun Fact: The summit of Mount Everest is located at an elevation of 8,848 meters (29,029 feet) above sea level.

16. The First Artificial Satellite (Sputnik 1)

In 1957, the Soviet Union launched Sputnik 1, the world's first artificial satellite. This event marked the beginning of the space race and sparked the development of modern satellite technology. Sputnik 1 orbited Earth for three months before re-entering the atmosphere and burning up.

Fun Fact: The launch of Sputnik 1 led to the formation of NASA in 1958, the United States' response to the Soviet space program.

17. Ancient Texts Uncovered: The Discovery of the Dead Sea Scrolls

In 1947, a Bedouin shepherd stumbled upon a remarkable discovery in caves near the Dead Sea: ancient scrolls containing some of the oldest known surviving manuscripts of the Hebrew Bible, as well as other Jewish texts. This find,

known as the Dead Sea Scrolls, shed invaluable light on early Judaism and the historical context of the Bible.
Fun Fact: The scrolls had been preserved for nearly 2,000 years in clay jars, hidden away in the dry desert climate.

18. The Creation of the Transcontinental Railroad

The completion of the Transcontinental Railroad in 1869 connected the East and West coasts of the United States. The railroad revolutionized transportation, allowing people and goods to travel across the country in days instead of months. This feat was a major driver of industrial expansion.

Fun Fact: The Transcontinental Railroad was constructed by thousands of workers, many of whom were Chinese immigrants and Irish laborers.

19. The First Successful Spacewalk

In 1965, Alexei Leonov, a Soviet cosmonaut, became the first person to conduct a spacewalk, or extravehicular activity (EVA). During his mission, he floated outside his spacecraft for 12 minutes, a groundbreaking achievement in space exploration.

Fun Fact: Leonov's spacewalk was an important step toward future missions, including those that landed on the Moon and conducted deep space exploration.

20. The Development of the Internet of Things (IoT)

The Internet of Things (IoT) refers to the network of physical objects embedded with sensors and software to connect and exchange data. This technology has rapidly advanced, with IoT applications in industries ranging from healthcare and agriculture to smart homes and transportation.

Ethereal Ray

Fun Fact: By 2025, it is estimated that there will be more than 75 billion connected IoT devices worldwide.

---

21. The Invention of the Refrigerator

In the early 19th century, the refrigerator was invented to preserve food by cooling it below ambient temperatures. The process of refrigeration revolutionized food storage, transportation, and preservation, and has become an essential part of modern living. Today, refrigeration is vital in homes, hospitals, and industries worldwide.

Fun Fact: The first refrigeration unit was created by Jacob Perkins in 1834, and it was powered by ether, a chemical that later gave way to more efficient refrigerants.

---

22. The Building of the Panama Canal

The Panama Canal is one of the most significant feats of engineering in history. Completed in 1914, it connects the Atlantic and Pacific Oceans, reducing the shipping distance by thousands of miles. The construction involved massive earthworks, overcoming disease, and solving complex engineering challenges.

Fun Fact: More than 75,000 workers were involved in the construction of the Panama Canal, with many of them suffering from yellow fever and malaria.

---

23. Ancient Engineering: Constructing the Colosseum

The Colosseum in Rome, completed in 80 AD, stands as a testament to the engineering prowess of the Roman Empire. This massive amphitheater, capable of holding 50,000 spectators, was used for gladiatorial contests, public spectacles, and even mock sea battles, demonstrating remarkable innovation in design and construction.

Fun Fact: The Colosseum's complex system of trap doors and elevators allowed for dramatic entrances of gladiators, animals, and scenery during events.

24. Ancient Grandeur: The Construction of the Colosseum

The Colosseum in Rome, completed in 80 AD, is one of the most iconic and enduring symbols of ancient Rome. This colossal amphitheater was used for gladiatorial games and public spectacles.

Fun Fact: The Colosseum could hold up to 50,000 spectators.

25. The Development of the Electric Motor

The electric motor converts electric energy into mechanical energy and is used in nearly every aspect of modern life. Developed over time by Michael Faraday, Nikola Tesla, and others, the electric motor powers everything from household appliances to industrial machines.

Fun Fact: The electric motor revolutionized the industrial world by providing efficient, reliable power for machines and vehicles.

26. The First Heart Transplant

In 1967, Christiaan Barnard performed the first successful human heart transplant in South Africa. This groundbreaking achievement marked a significant milestone in medical history, offering new hope for patients with end-stage heart disease.

Fun Fact: The first heart transplant patient, Louis Washkansky, lived for 18 days after the operation.

---

27. Diving Deep: The Invention of the Bathyscaphe

The bathyscaphe, a type of deep-sea submersible, was invented by Auguste Piccard in the mid-20th century. These vessels allowed humans to explore the deepest parts of the ocean for the first time, leading to incredible discoveries about marine life and the ocean floor.

Fun Fact: In 1960, the bathyscaphe *Trieste* descended to the bottom of the Mariana Trench, the deepest part of the ocean, nearly 11 kilometers (7 miles) below the surface.

28. Picture This: The Invention of the Polaroid Camera

Edwin Land's invention of the Polaroid camera in 1948 revolutionized photography by making instant photography possible. This breakthrough allowed people to see their photos within minutes of taking them, transforming the way people captured and shared memories.

Fun Fact: The Polaroid camera's instant film technology was a closely guarded secret for many years.

29. Power in Hand: Bell's Telephone

Alexander Graham Bell's invention of the telephone in 1876 revolutionized communication, enabling real-time voice conversations over long distances. This invention dramatically changed personal and business interactions.

Fun Fact: Bell's initial vision for the telephone was for it to be used to transmit musical performances.

30. Beyond the Obvious: The Invention of the Computer

The invention of the computer, a process involving numerous innovators over many decades, has fundamentally reshaped society. From the room-sized ENIAC of the 1940s to today's powerful microprocessors, computers have revolutionized information processing.

Fun Fact: The theoretical foundation for modern computers was laid by Alan Turing in the 1930s.

31. Enduring Structure: The Great Wall of China

The Great Wall of China, one of the most impressive architectural feats in history, is a testament to the ingenuity and determination of the ancient Chinese. Built over centuries, this immense structure served as a defensive barrier.

Fun Fact: Contrary to popular belief, the Great Wall is not visible from the Moon with the naked eye.

32. A World Connected: The Creation of the Internet

The creation of the internet, a network connecting billions of devices worldwide, has transformed communication, commerce, and access to information. What began as a U.S.

Department of Defense project (ARPANET) in the 1960s evolved into the global network we rely on today.

Fun Fact: The "@" symbol, now ubiquitous in email addresses, was chosen by Ray Tomlinson in 1971.

33. Reaching the Roof of the World: The First Everest Climb

Reaching the summit of Mount Everest, the world's highest peak, was a seemingly impossible dream until 1953, when Sir Edmund Hillary and Tenzing Norgay achieved this historic feat. Their accomplishment demonstrated the extraordinary power of human endurance.

Fun Fact: Hillary and Norgay had to battle extreme altitude, freezing temperatures, and treacherous terrain.

34. Transforming Textiles: The Cotton Gin's Impact

Eli Whitney's invention of the cotton gin in 1793 was a pivotal moment in agricultural history. By automating the process of separating cotton fibers from seeds, the cotton gin dramatically increased production.

Fun Fact: The cotton gin, while increasing productivity, also, unfortunately, led to an increase in the demand for slave labor.

35. Medical Breakthrough: The Development of Blood Transfusions

The development of safe and effective blood transfusions in the early 20th century was a medical revolution. The discovery of blood types by Karl Landsteiner in 1901 was a crucial step, allowing doctors to match donors and recipients.

Fun Fact: Blood transfusions played a critical role in saving lives during World War I.

36. A Nation United: The Transcontinental Railroad

The completion of the Transcontinental Railroad in 1869, linking the eastern and western coasts of the United States, was a monumental engineering achievement. This new transportation link dramatically reduced travel time across the vast nation.

Fun Fact: The "Golden Spike" ceremony, where the final spike was driven to complete the railroad, was one of the first national media events.

37. Revolutionizing Surgery: The First Heart-Lung Machine

The invention of the heart-lung machine in 1953 by John Gibbon was a breakthrough that made open-heart surgery possible. This complex device temporarily takes over the functions of the heart and lungs.

Fun Fact: Gibbon's first successful use of the heart-lung machine on a human patient was a milestone in cardiac surgery.

38. Exploring the Depths: The First Powered Submarine

The development of the first practical powered submarines in the late 19th and early 20th centuries marked a new era in naval warfare and underwater exploration. Inventors like John Holland and Simon Lake pioneered designs.

Fun Fact: Early submarines were cramped, dangerous, and often unreliable.

39. Engine-less Flight: The Feat of Human-Powered Aircraft

The dream of flying like a bird without the aid of an engine has long captivated inventors. While challenging, human-powered aircraft have achieved remarkable feats, demonstrating the power of ingenuity.

Fun Fact: The Gossamer Albatross, a human-powered aircraft, successfully crossed the English Channel in 1979.

40. Unlocking the Atom: The Manhattan Project

The Manhattan Project, a research and development undertaking during World War II, produced the first nuclear weapons. This massive scientific and engineering effort involved the collaboration of thousands of scientists and engineers.

Fun Fact: The Manhattan Project was conducted in such secrecy that even Vice President Truman didn't know about it until he became president.

Conclusion:

Human ingenuity knows no bounds. From the first powered flight to building vast engineering marvels, humanity's ability to overcome challenges, innovate, and push the boundaries of what is possible is truly remarkable.

These feats are a testament to our creativity, perseverance, and the pursuit of progress. Each one has helped shape the modern world, opening up new opportunities and inspiring future generations to continue exploring the limits of human potential.

Chapter 10: Incredible Animals

Animals have always fascinated humans, from the majestic creatures that roam the savannah to the mysterious beings that lurk in the depths of the oceans.

Throughout history, people have marveled at the diversity and complexity of animal life, discovering amazing facts about how they survive, thrive, and adapt to their environments.

This chapter will take you on a journey through the animal kingdom, highlighting some of the most **extraordinary creatures** and **incredible facts** about the living world.

1. Nature's Giants: The Blue Whale

The blue whale is the largest animal to have ever lived on Earth, growing up to 100 feet long and weighing as much as 200 tons. Despite their enormous size, these gentle giants feed on tiny shrimp-like creatures called krill.

Fun Fact: A blue whale's heart can weigh as much as a small car!

2. Speed Demons: The Cheetah

The cheetah is capable of reaching speeds up to 60 miles per hour (97 km/h) in short bursts, making it the fastest land animal. Their incredible speed is a result of their specialized bodies.

Fun Fact: A cheetah's acceleration is faster than most sports cars.

3. Marathon Migrators: The Monarch Butterfly

The monarch butterfly is famous for its long-distance migration, traveling up to 3,000 miles from Canada and the United States to central Mexico. During migration, monarchs rely on natural cues like the sun.

Fun Fact: Monarch butterflies are the only insects known to make a two-way migration.

4. Gentle Giants: The Elephant

Elephants are the largest land mammals, known for their intelligence, emotional depth, and strong social bonds. They can communicate over long distances using low-frequency sounds.

Fun Fact: Elephants have a highly developed sense of smell.

5. Masters of Disguise: The Octopus

The octopus is an incredibly intelligent and adaptable creature, capable of changing its skin color and texture to blend in with its environment. This ability helps it avoid predators and catch prey.

Fun Fact: Octopuses can solve puzzles, open jars, and even escape from sealed tanks.

6. Long-Distance Fliers: The Arctic Tern

The Arctic tern holds the record for the longest migration of any animal, traveling up to 25,000 miles each year from its breeding grounds in the Arctic to the Antarctic and back.

Fun Fact: The Arctic tern's migration spans three continents.

7. Regeneration Champions: The Axolotl

The axolotl, a type of salamander, has the amazing ability to regenerate lost limbs, spinal cord, and even parts of its heart and brain. Unlike most other animals, this fascinating creature remains in its larval stage.

Fun Fact: The axolotl is often called the "Mexican walking fish".

8. Lizard Kings: The Komodo Dragon

The Komodo dragon is the largest living species of lizard, growing up to 10 feet long and weighing as much as 150 pounds. It is a powerful predator, using its sharp claws and strong jaws to hunt large prey.

Fun Fact: The Komodo dragon has bacteria in its saliva that help it kill prey.

9. Underwater Gunners: The Pistol Shrimp

The pistol shrimp is known for its powerful claws that snap shut so quickly they create a bubble that stuns prey. This snap also generates heat, light, and a shockwave.

Fun Fact: The pistol shrimp's claw snap is so loud that it can be heard by humans above the water.

10. Pollinator Powerhouse: The Honeybee

Honeybees are essential for pollinating plants, helping to create fruits, vegetables, and flowers. A single bee can visit 5,000 flowers in one day, and a colony of bees can pollinate an entire orchard.

Fun Fact: Honeybees communicate with each other through a dance.

11. Slow-Motion Marvels: The Sloth

Sloths are incredibly slow-moving animals, spending up to 20 hours a day resting or hanging upside down in the trees of Central and South America. Their slow metabolism and low-energy diet are the key to their lifestyle.

Fun Fact: Sloths only defecate once a week.

12. Scaly Defenders: The Pangolin

The pangolin is the only mammal entirely covered in scales made of keratin, the same substance as human nails. When threatened, it rolls into a tight ball, using its sharp claws to defend itself.

Fun Fact: Pangolins are often called "the world's most trafficked mammal".

13. Towering Giants: The Giraffe

The giraffe is the tallest land animal, reaching heights of up to 18 feet. Its long neck helps it reach leaves high in trees, but it also has a unique circulatory system.

Fun Fact: Despite their long necks, giraffes have only seven vertebrae in their necks, the same number as humans!

14. Unique Caretakers: The Male Seahorse

In the animal kingdom, it's the male seahorse that carries and nurtures the eggs. The female seahorse deposits her eggs into the male's brood pouch, where he fertilizes them and carries them until they hatch.

Fun Fact: The male seahorse gives birth to fully formed baby seahorses!

15. Apex Predators: The Great White Shark

The great white shark is one of the most feared predators of the ocean. These incredible hunters use their keen sense of smell and powerful jaws to catch prey.

Fun Fact: Great white sharks can swim at speeds of up to 35 mph.

16. Sleepy Tree-Dwellers: The Koala

Koalas are marsupials native to Australia, known for spending up to 20 hours a day sleeping in eucalyptus trees. Their diet consists almost entirely of eucalyptus leaves, which provide little energy.

Fun Fact: Despite their cuddly appearance, koalas have sharp claws and teeth.

17. Silent Hunters: The Snow Leopard

The snow leopard is a rare, elusive cat native to the mountains of Central Asia. Known for its beautiful spotted coat and incredible ability to adapt to cold climates, this predator stalks its prey in the rocky, snowy terrain.

Fun Fact: Snow leopards are known for their powerful hind legs.

18. Sea Unicorns: The Narwhal

The narwhal, often called the "unicorn of the sea," is known for its long, spiral tusk that can grow up to 10 feet in length. The tusk is actually an elongated tooth.

Fun Fact: Narwhals are often spotted in Arctic waters.

19. Arctic Royalty: The Polar Bear

The polar bear is the largest bear species, and it is perfectly adapted to the harsh conditions of the Arctic. Its thick fur, powerful body, and sharp claws help it hunt for seals.

Fun Fact: Polar bears have black skin under their white fur.

20. Nature's Oddity: The Platypus

The platypus is one of the most unusual animals on Earth, as it is a mammal that lays eggs, has webbed feet, and can produce venom. Native to Australia, it's a living example of nature's creativity.

Fun Fact: The platypus's venom is powerful enough to cause severe pain in humans.

21. Wide-Eyed Hunters: The Hammerhead Shark

The hammerhead shark is known for its distinctive, wide head, which gives it a 360-degree view of its surroundings. This adaptation helps it locate prey.

Fun Fact: Hammerhead sharks have specialized electroreceptors in their heads.

22. Gentle Giants: The Gorilla

The gorilla is the largest of the primates and known for its strength, intelligence, and social structure. Despite their massive size, gorillas are peaceful creatures and live in tight-knit family groups.

Fun Fact: Gorillas share about 98% of their DNA with humans.

23. Cold Survivors: The Arctic Fox

The Arctic fox is uniquely adapted to survive in some of the coldest environments on Earth. With thick fur and a bushy tail, this small mammal can withstand temperatures as low as -58°F.

Fun Fact: The Arctic fox changes the color of its fur from brown in the summer to white in the winter.

24. Deep-Sea Hunters: The Anglerfish

The anglerfish is one of the most unique fish in the ocean. It has a bioluminescent lure on top of its head, which it uses to attract prey in the dark depths of the ocean.

Fun Fact: The anglerfish's lure contains bacteria that produce light.

25. Burrowing Experts: The Wombat

Native to Australia, the wombat is a burrowing marsupial known for its square-shaped poop, which helps prevent it from rolling away. Wombats dig complex burrows to escape the extreme heat and predators.

Fun Fact: Wombats are one of the few animals with square-shaped poop, which is thought to help mark their territory.

26. Colorful Killers: The Poison Dart Frog

The poison dart frog is a small, brightly colored frog found in the rainforests of Central and South America. These frogs get their name from the potent toxins in their skin, which serve as a defense mechanism against predators.

Fun Fact: Poison dart frogs' bright colors are a warning sign to predators that they are toxic and not safe to eat.

27. Deadly Serpents: The King Cobra

The king cobra is the world's longest venomous snake, reaching lengths of up to 18 feet. It can strike with deadly precision, delivering a potent venom that can kill an elephant within hours.

Fun Fact: The king cobra's venom can paralyze its prey, but it usually prefers to avoid humans unless provoked.

28. Tree-Dwellers: The Red Panda

The red panda, also known as the fire fox, is a small mammal native to the mountain forests of Himalayas and China. Despite its name, it is not closely related to the giant panda and is known for its solitary nature and tree-dwelling habits.

Fun Fact: Red pandas use their bushy tails for balance and warmth, curling them around themselves when they sleep.

29. Fierce Carnivores: The Tasmanian Devil

The Tasmanian devil is a small but fierce carnivorous marsupial native to Tasmania. Known for its loud screeches and fierce temperament, the Tasmanian devil is a scavenger that feeds on anything from small mammals to carrion.

Fun Fact: Tasmanian devils have the strongest bite of any carnivorous marsupial and can crush bone with their powerful jaws.

30. Sea Cows: The Manatee

Manatees, often referred to as sea cows, are slow-moving marine mammals that feed on seagrasses in shallow coastal waters. Despite their gentle and slow demeanor, these creatures are surprisingly agile and can travel long distances.

Fun Fact: Manatees are herbivores and spend most of their time grazing on aquatic plants. They can eat up to 10% of their body weight in plants daily.

31. Speed Demons: The Peregrine Falcon

The peregrine falcon is the fastest bird in the world, capable of reaching speeds over 240 mph during its hunting dives, making it one of the fastest animals on the planet. They are powerful hunters, preying on smaller birds like pigeons.

Fun Fact: The peregrine falcon's diving speed is faster than the speed of sound in air!

32. River Giants: The Hippopotamus

The hippopotamus may look docile, but it is one of the most dangerous animals in Africa. Weighing up to 4,000 pounds, these herbivores can run surprisingly fast on land and are highly territorial in water.

Fun Fact: Despite being herbivores, hippopotamuses are responsible for more human deaths in Africa than any other large animal.

33. Color-Changers: The Chameleon

Chameleons are known for their ability to change color, which helps them blend into their environment or communicate with other chameleons. This process is controlled by special pigment cells in their skin called chromatophores.

Fun Fact: Chameleons change color based on their mood, temperature, and even social interactions, not just for camouflage.

34. Powerful Predators: The Jaguar

The jaguar is the largest cat in the Americas and is known for its incredible strength and powerful bite. Jaguars can crush the skulls or shells of their prey with ease, often taking down animals as large as caimans or deer.

Fun Fact: Jaguars are the only big cats that regularly hunt crocodiles and caimans by biting through their skulls.

35. Gentle Giants of the Sea: The Whale Shark

The whale shark is the largest fish in the world, growing up to 40 feet long. Despite its enormous size, the whale shark feeds on tiny plankton and small fish, making it a filter feeder.

Fun Fact: Whale sharks have more than 300 teeth, but they don't use them to eat. They filter-feed through their gills instead!

36. Master Mimics: The Lyrebird

The lyrebird, native to Australia, is famous for its incredible ability to mimic sounds. It can imitate chainsaws, car alarms, and other animals, making it one of the best vocal mimics in the animal kingdom.

Fun Fact: A lyrebird's mimicry can be so accurate that it can imitate human-made sounds, like camera shutters and even mobile phone rings!

37. Clever Foragers: The Raccoon

Raccoons are known for their dexterity and problem-solving abilities. With their front paws almost like human hands, they are adept at opening containers, doors, and latches. They are also very curious and can get into all kinds of mischief.

Fun Fact: Raccoons have been known to open locks, open doors, and even learn how to use tools!

38. Tool-Users: The Sea Otter

Sea otters are remarkable marine mammals known for using stones and other tools to crack open shellfish. They are one of the few animal species to use tools, demonstrating both intelligence and dexterity.

Fun Fact: Sea otters have a pouch under their forearms where they store their favorite rocks for cracking open shellfish.

39. Underwater Superheroes: The Mantis Shrimp

The mantis shrimp has one of the most powerful strikes in the animal kingdom. Its club-like appendages can strike with the speed of a bullet, stunning or killing prey with a single blow.

Fun Fact: The mantis shrimp can strike its prey with speeds up to 80 kilometers per hour, and the impact generates light and heat!

40. Nighttime Navigators: The Bat

Bats are the only mammals capable of true flight. Using echolocation, they navigate through the night to catch insects or even small vertebrates. Bats play a critical role in controlling insect populations.

Fun Fact: Bats are the only mammals that can truly fly and are capable of eating up to half their body weight in insects each night.

Conclusion:

The world of **animals** is full of wonder and diversity, from the smallest insects to the most powerful predators. Each species has evolved in remarkable ways to survive and thrive in its environment.

As we continue to explore the wonders of the animal kingdom, it's essential that we appreciate and protect these incredible creatures that make our planet so unique. Whether through their extraordinary adaptations, their amazing abilities, or their incredible migrations, animals remind us of the beauty and complexity of life on Earth.

The Ultimate Rare and Unusual Knowledge Quiz

Instructions: Put your thinking caps on! Choose the best answer for each question. Good luck!

Chapter 1: History's Hidden Gems

1. Which two U.S. Presidents share a series of coincidences, including being elected 100 years apart?
A. Washington and Jefferson
B. Lincoln and Kennedy
C. Roosevelt and Eisenhower
D. Adams and Jackson

2. What was the title of the novella written by Morgan Robertson that eerily predicted the Titanic disaster?
A. *The Wreck of the Titan*
B. *Futility*
C. *Unsinkable*
D. *A Night to Remember*

3. What astronomical event coincided with both the birth and death of Mark Twain?
A. The appearance of Halley's Comet
B. A total solar eclipse

C. The Leonid meteor shower
D. A supermoon

4. What historical artifact contains three scripts and helped scholars decipher Egyptian hieroglyphs?
A. The Dead Sea Scrolls
B. The Rosetta Stone
C. The Code of Hammurabi
D. The Sumerian Tablets

5. Who was the first emperor of China, famous for unifying the country and constructing the Great Wall?
A. Sun Tzu
B. Qin Shi Huang
C. Kublai Khan
D. Confucius

Chapter 2: Unbelievable Science

6. What shape is the massive storm at Saturn's north pole?
A. Circular
B. Square
C. Hexagonal
D. Spiral

7. Which jellyfish species is considered biologically immortal?
A. Aurelia aurita
B. Turritopsis dohrnii
C. Chrysaora quinquecirrha
D. Cyanea capillata

8. What is the Golden Record on the Voyager probes?
A. A map of the solar system

B. A message to alien civilizations
C. A device for measuring radiation
D. A sample of Earth's atmosphere

9. What subatomic particle can pass through solid matter undetected?
A. Quark
B. Electron
C. Neutrino
D. Proton

10. What force prevents Earth's atmosphere from drifting into space?
A. Magnetic Fields
B. Gravity
C. Dark Matter
D. Solar Winds

Chapter 3: Technology That Changed the World

11. What was Alexander Graham Bell's famous first message transmitted via telephone?
A. "Can you hear me now?"
B. "Watson, come here!"
C. "Testing, one, two, three."
D. "Hello, world!"

12. Who coined the term *Artificial Intelligence*?
A. Alan Turing
B. John McCarthy
C. Isaac Asimov
D. Ada Lovelace

13. What was the original price of the Ford Model T car?
A. $100
B. $250
C. $850
D. $3,000

14. What major medical breakthrough was developed using mold in 1928?
A. Insulin
B. Penicillin
C. Polio Vaccine
D. Smallpox Vaccine

15. What computing advancement allows machines to process data like a human brain?
A. Quantum Memory
B. Artificial Neural Networks
C. Holographic Processing
D. Neural Cloning

Chapter 4: Cultural Curiosities

16. What is the purpose of a dreamcatcher in Native American culture?
A. To catch food
B. To ward off evil spirits
C. To protect from bad dreams
D. To tell stories

17. What was the only clue left behind by the vanished Roanoke Colony?
A. A map
B. The word "CROATOAN"

C. A buried treasure
D. A skeleton

18. Who first described the lost city of Atlantis?
A. Homer
B. Plato
C. Aristotle
D. Herodotus

19. What mysterious giant figures are etched into the desert in Peru?
A. Stonehenge
B. The Moai Statues
C. The Nazca Lines
D. The Sphinx of Giza

20. What is the ancient Japanese art of folding paper called?
A. Haiku
B. Katana
C. Origami
D. Ikebana

Chapter 5: Mysteries and Unsolved Puzzles

21. What is unusual about the Piri Reis Map?
A. It's drawn on a turtle shell
B. It depicts Antarctica
C. It's written in invisible ink
D. It's the oldest map in the world

22. What is the Voynich Manuscript?
A. A medieval cookbook
B. A book written in an undecipherable script

C. A guide to alchemy
D. A collection of love poems

23. What is the Tunguska Event?
A. A volcanic eruption
B. A massive explosion in Siberia
C. A political revolution
D. A major earthquake

24. What historical mystery involves a prisoner forced to wear a mask?
A. The Man in the Iron Mask
B. The Phantom of the Opera
C. The Black Hooded Monk
D. The Scarlet Prisoner

25. What famous cryptographic artifact has never been deciphered?
A. The Phaistos Disk
B. The Emerald Tablet
C. The Rosetta Stone
D. The Voynich Manuscript

Chapter 6: Unexpected Natural Phenomena

26. What rare phenomenon involves glowing orbs appearing during thunderstorms?
A. Plasma clouds
B. Sprites
C. Ball lightning
D. Static flares

Ethereal Ray

27. What natural event causes fish or frogs to fall from the sky?
A. Tornado rain
B. Animal storm
C. Waterspouts lifting them
D. Aquatic precipitation

28. What eerie cloud formation appears as a rolling tube stretching across the sky?
A. Shelf cloud
B. Lenticular cloud
C. Morning Glory cloud
D. Stratospheric wave

Chapter 6: Unexpected Natural Phenomena

29. What is the longest continuous lightning storm on Earth?
A. The Sahara Flash
B. The Catatumbo Lightning
C. The Eternal Thunderstorm
D. The Maracaibo Sparks

30. What natural phenomenon causes ocean waves to glow blue at night?
A. Deep-sea reflection
B. Bioluminescent plankton
C. Algae combustion
D. Electromagnetic interference

31. What mysterious geological formation features large stones that move on their own?
A. The Sliding Rocks of Death Valley
B. The Floating Stones of Easter Island

C. The Shifting Stones of Patagonia
D. The Phantom Rocks of Mongolia

32. Which location is known for its eerie "singing" sand dunes?
A. The Sahara Desert
B. The Namib Desert
C. The Gobi Desert
D. The Atacama Desert

33. What is the name of the strange red-colored waterfall in Antarctica?
A. The Ruby Falls
B. The Blood Falls
C. The Crimson Cascade
D. The Iron Stream

Chapter 7: History's Greatest Conspiracies

34. What organization is suspected to have hidden the lost treasure of the Knights Templar?
A. The Illuminati
B. The Freemasons
C. The Vatican
D. The Skull and Bones Society

35. What ancient structure is believed to contain secret chambers yet to be discovered?
A. The Colosseum
B. The Great Pyramid of Giza
C. Machu Picchu
D. Stonehenge

36. What mystery surrounds the Dyatlov Pass Incident?
A. A UFO sighting
B. Unexplained deaths of hikers
C. The discovery of an unknown language
D. A village that vanished overnight

37. Which famous historical figure is rumored to have faked their death?
A. Napoleon Bonaparte
B. Adolf Hitler
C. Genghis Khan
D. Leonardo da Vinci

38. What missing document is believed to contain hidden knowledge about humanity?
A. The Book of Enoch
B. The Lost Gospel of Judas
C. The Dead Sea Scrolls
D. The Vatican's Hidden Texts

39. What mysterious island was once listed on maps but later disappeared?
A. Atlantis
B. Hy-Brasil
C. Avalon
D. Bermeja

40. What secret society was rumored to control world events from the shadows?
A. The Rosicrucians
B. The Order of the Phoenix
C. The Golden Dawn
D. The Black Cross

Chapter 8: Breakthroughs That Changed the World

41. What revolutionary technology allows scientists to edit DNA with precision?
A. CRISPR
B. GeneScape
C. RNA Splicing
D. DNA Sculpting

42. What was the first planet discovered outside our solar system?
A. Proxima Centauri b
B. 51 Pegasi b
C. Kepler-22b
D. Trappist-1d

43. What global network of telescopes captured the first image of a black hole?
A. The Keck Observatory
B. The Event Horizon Telescope
C. The Hubble Network
D. The Supermassive Tracker

44. What subatomic particle, discovered in 2012, helps explain how objects gain mass?
A. Neutrino
B. Graviton
C. Higgs Boson
D. Tachyon

45. What innovative vaccine technology was widely used for the first time in 2020?
A. DNA-based vaccine

B. mRNA vaccine
C. Plasma-based vaccine
D. Virus capsule vaccine

46. What scientific breakthrough confirmed that the universe is expanding?
A. The Large Hadron Collider Test
B. The Cosmic Inflation Model
C. Hubble's Redshift Observations
D. The Dark Matter Simulation

47. What futuristic energy source could produce clean, limitless power by mimicking the Sun?
A. Fission Reactors
B. Biofuel Harvesting
C. Nuclear Fusion
D. Cold Plasma Generators

48. What type of rocket innovation made space travel more affordable?
A. Disposable Boosters
B. Reusable Rockets
C. Mini Propulsion Capsules
D. Ion Thrusters

49. What material is being developed to replace plastic with biodegradable alternatives?
A. Carbon Fiber Polymers
B. Bacterial Cellulose
C. Hydrogel Plastics
D. Silicon Biowaste

50. What experiment confirmed the existence of gravitational waves, proving Einstein's theory?

A. The LIGO Experiment
B. The Higgs Field Test
C. The CERN Supercollider
D. The Hawking Wave Analysis

Answer Key:

1. B – Lincoln and Kennedy
2. B – *Futility*
3. A – The appearance of Halley's Comet
4. B – The Rosetta Stone
5. B – Qin Shi Huang
6. C – Hexagonal
7. B – Turritopsis dohrnii
8. B – A message to alien civilizations
9. C – Neutrino
10. B – Gravity
11. B – "Watson, come here!"
12. B – John McCarthy
13. C – $850
14. B – Penicillin
15. B – Artificial Neural Networks
16. C – To protect from bad dreams
17. B – The word "CROATOAN"
18. B – Plato
19. C – The Nazca Lines
20. C – Origami
21. B – It depicts Antarctica
22. B – A book written in an undecipherable script
23. B – A massive explosion in Siberia
24. A – The Man in the Iron Mask
25. A – The Phaistos Disk

26. C – Ball lightning
27. C – Waterspouts lifting them
28. C – Morning Glory cloud
29. B – The Catatumbo Lightning
30. B – Bioluminescent plankton
31. A – The Sliding Rocks of Death Valley
32. B – The Namib Desert
33. B – The Blood Falls
34. B – The Freemasons
35. B – The Great Pyramid of Giza
36. B – Unexplained deaths of hikers
37. B – Adolf Hitler
38. D – The Vatican's Hidden Texts
39. B – Hy-Brasil
40. A – The Rosicrucians
41. A – CRISPR
42. B – 51 Pegasi b
43. B – The Event Horizon Telescope
44. C – Higgs Boson
45. B – mRNA vaccine
46. C – Hubble's Redshift Observations
47. C – Nuclear Fusion
48. B – Reusable Rockets
49. B – Bacterial Cellulose
50. A – The LIGO Experiment

Review Request
from Professor Atlas – Share Your Discovery!

Dear Reader,

Our journey through **rare and unusual knowledge** has come to an end, but the adventure doesn't have to stop here! If you enjoyed this book, **I'd love to hear from** you.

Leave a review on Amazon - Your feedback helps others discover this book and fuels future explorations!

Snap a photo or record a short video of your favorite fact or chapter and upload it with your Amazon review! Whether it's a mind-blowing discovery, a surprising historical mystery, or an incredible scientific breakthrough, share what fascinated you the most.

Your insights make a difference, and who knows? Your review might even inspire the next volume of *The Ultimate Book of Rare and Unusual Knowledge!*

Thank you for being part of this journey—stay curious, keep exploring, and never stop learning!

— **Professor Atlas**

Check my other ongoing books in the series!

Check my growing book series!

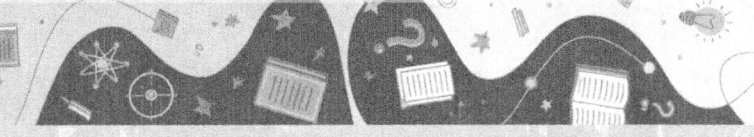

ELEVEN BOOKS TO EXPLORE

Scan the QR below